Properties, Powers and Str

Routledge Studies in Metaphysics

Properties, Powers and Structures

Issues in the Metaphysics of Realism

**Edited by Alexander Bird,
Brian Ellis, and Howard Sankey**

Routledge
Taylor & Francis Group

NEW YORK   LONDON

First published 2012
by Routledge

2 Park Square, Milton Park, Abingdon, Oxon OX14 4RN
711 Third Avenue, New York, NY 10017, USA

Routledge is an imprint of the Taylor & Francis Group, an informa business

First issued in paperback 2016

Typeset in Sabon by IBT Global.

Library of Congress Cataloging-in-Publication Data
Properties, powers, and structures : issues in the metaphysics of realism /
    edited by Alexander Bird, Brian Ellis, and Howard Sankey.
        p. cm. — (Routledge studies in metaphysics ; 5)
    Proceedings of a conference held in July 2009 at the University of
Melbourne.
    Includes bibliographical references (p.   ) and index.
    1. Realism—Congresses.   2. Science—Philosophy—Congresses.
I. Bird, Alexander, 1964–   II. Ellis, B. D. (Brian David), 1929–
III. Sankey, Howard.
    Q175.32.R42P74 2012
    149'.2—dc23
    2011041091

ISBN 13: 978-0-415-89535-4 (hbk)
ISBN 13: 978-1-138-24528-0 (pbk)

CONTENTS

NOTES ON CONTRIBUTORS

Rani Lill Anjum is Research Fellow of Philosophy and Director of the cross-disciplinary research project CauSci—Causation in Science—at the Norwegian University of Life Sciences (UMB). She is co-author of *Getting Causes From Powers* (Oxford University Press 2011) and a number of articles on causation with Stephen Mumford.

D. M. Armstrong is Emeritus Professor of Philosophy at Sydney University. His short book *Sketch for a Systematic Metaphysics* was published in 2010 by Oxford University Press.

William A. Bauer is a Teaching Assistant Professor of Philosophy at North Carolina State University. His research addresses problems in metaphysics and the philosophy of science, and focuses especially on dispositional properties and causal powers. His current and future research includes projects on emergent properties, the dispositional nature of beliefs, and metaphysical problems that arise in bioethics. He has published articles in *The Reasoner* and *Erkenntnis*.

Alexander Bird is Professor of Philosophy at the University of Bristol. He was Principal Investigator on the Arts and Humanities Research Council's Metaphysics of Science project (2006–10). He is author of *Nature's Metaphysics: Laws and properties* (Oxford 2007) and was co-editor of the *British Journal for the Philosophy of Science* (2004–11).

Corinne L. Bloch is a Ph.D. candidate in Philosophy at the Cohn Institute for the History and Philosophy of Science and Ideas, Tel Aviv University, as well as a Ph.D. candidate in Animal Science, specializing in neuro-endocrinology, at the Hebrew University of Jerusalem. She is currently a visiting scholar at the History and Philosophy of Science department at the University of Pittsburgh. She has forthcoming papers on scientific concepts and definitions.

Brian Ellis is Emeritus Professor of Philosophy at La Trobe University and Professorial Fellow in Philosophy at the University of Melbourne. He is the author of six books on metaphysics or philosophy of science, including

The Metaphysics of Scientific Realism (Acumen 2009). He was Editor of the *Australasian Journal of Philosophy* (1978–89).

Sharon R. Ford received her Ph.D. in Philosophy at the University of Queensland, Australia. Her research and publishing interests currently focus upon dispositions and powers, continuity and continuums, Power Holism, and Priority Monism. Wider interests include theory of properties, metaphysics of science, and philosophy of space and time.

James Franklin is a professor of mathematics at the University of New South Wales. He is the author of *Corrupting the Youth: A History of Philosophy in Australia* and of the chapter on 'Aristotelian realism' in the 'Philosophy of Mathematics' volume of the North-Holland Elsevier *Handbook of the Philosophy of Science.*

Max Kistler is Professor in the Department of Philosophy at the University of Paris 1 (Panthéon-Sorbonne). He is author of *Causation and Laws of Nature* (Routledge 2006), and has edited (with B. Gnassounou) *Dispositions and Causal Powers* (Ashgate 2007), and special issues of *Synthese* **151** (2006), on 'New Perspectives on Reduction and Emergence in Physics, Biology, and Psychology', and *Philosophical Psychology* **22** (2009), on 'Cognition and Neurophysiology: Mechanism, Reduction, and Pluralism'.

Patrick McGivern is a Lecturer in Philosophy at the University of Wollongong in New South Wales. His primary research interests are in philosophy of science and philosophy of mind. He is especially interested in physics-based accounts of reduction, emergence, and inter-theoretic explanation, and in how such accounts can be applied across different sciences.

Stephen Mumford is Professor of Metaphysics at the University of Nottingham and Professor II at the Norwegian University of Life Science (UMB). He is author of *Dispositions* (Oxford 1998), *Laws in Nature* (Routledge 2004), *David Armstrong* (Acumen 2007), *Watching Sport: Aesthetics, Ethics and Emotion* (Routledge 2011) and co-author of *Getting Causes from Powers*, with Rani Lill Anjum (Oxford 2011) as well as editor of *Russell on Metaphysics* (Routledge 2003) and George Molnar's posthumous book *Powers* (Oxford 2003).

Anne Newstead is a Research Associate in the School of Mathematics and Statistics at the University of New South Wales.

Stathis Psillos is Professor of Philosophy of Science and Metaphysics at the University of Athens, Greece. He is the author of: *Knowing the Structure of Nature* (Palgrave 2009), *Philosophy of Science A–Z* (Edinburgh University Press 2007); *Causation and Explanation* (McGill–Queens U.P. 2002); and

Scientific Realism: How Science Tracks Truth (Routledge 1999). He is also the co-editor of *The Routledge Companion to Philosophy of Science* (Routledge 2008). He has published more than seventy papers in learned journals and books on scientific realism, causation, explanation, and the history of philosophy of science. He has served as the President of the European Philosophy of Science Association (2007–9) and is currently co-editor of *Metascience*.

Howard Sankey is Associate Professor in the School of Historical and Philosophical Studies at the University of Melbourne, where he teaches into the programmes of Philosophy and History and Philosophy of Science. He has published on topics in the general philosophy of science, such as incommensurability, rational theory choice, relativism, method and scientific realism. He is currently working on a project about the relationship between science and common sense, and the relationship between scepticism and epistemic relativism.

Emma Tobin is Lecturer in the Philosophy of Science in the Department of Science and Technology Studies at University College London. Her main research interests are in metaphysics of science and philosophy of chemistry. She is the author of 'Microstructuralism and Macromolecules', published in *Foundations of Chemistry* in 2009, and also 'Crosscutting Natural Kinds and the Hierarchy Thesis' in *The Semantics and Metaphysics of Natural Kinds* (Routledge 2010), edited by Helen Beebee and Nigel Leary.

Matthew Tugby obtained his Ph.D. at the University of Nottingham in 2010 under the UK Arts and Humanities Research Council's Metaphysics of Science project. He has since published on the topic of dispositions and is now a Teaching Fellow at the University of Birmingham.

Barbara Vetter is a junior professor at Humboldt-Universität, Berlin. She is currently working on a dispositional theory of metaphysical modality.

INTRODUCTION

Brian Ellis, Howard Sankey, and Alexander Bird

Metaphysical realism is a view about the nature of reality that derives from science. It presupposes scientific realism, and seeks to construct a theory about the nature of reality based on this assumption. A principle that guides much of metaphysical realism is that first elaborated by John Fox (1987) in his 'Truthmaker' theory. It is the claim that what is true depends ontologically on what there is, i.e. that truth supervenes on being. Consequently, the task of constructing a metaphysics for scientific realism is seen as being that of providing an adequate set of truthmakers for the established truths of science and mathematics. It must begin with the science, draw conclusions about what truths need to be explained ontologically, and end with the required explanations.

Metaphysical realism seems set to become the dominant metaphysics of our time. It is, of course, anti-positivist, and also anti-nominalist and anti-reductionist. But it is not a reversion to the kind of idealism that was common in scientific philosophy in the nineteenth century. Nor is it driven by the demands of logic or language, as so much of twentieth century philosophy of science evidently was. For its focus is on the nature of reality, given a realistic understanding of modern science. It is not about explaining away what appear to be the realistic implications of science. Nor is it about how to express the results of science in an extensional language. And, gone are the days when it could seriously be said that 'to be is to be the value of a variable'. For the general view now is that neither logic nor language has much to tell us about what there is. Logic is seen as being independent of it, and if science makes extensive use of modal languages, then this establishes a prima facie case for seeking ways of identifying the truthmakers for the propositions of such languages. The general attitude is that to find out about reality, and hence to construct an adequate theory of it, we have to start by considering what our best science has to tell us. And, in the spirit of scientific realism, we should take this seriously, and not try to mould it to squeeze reality into any pre-existing framework of logic or language.

Nor should we be quick to reject any metaphysical position just because it implies that the world is not as we intuitively imagine it to be. For metaphysics, as metaphysical realists understand it, is not an a priori discipline, and our metaphysical intuitions, such as they are, are likely to have been fashioned by the great metaphysical traditions of the past. Plato, Aristotle, Descartes, Locke, Hume and Kant were all, undoubtedly, great and influential metaphysicians. But they were all responding to the sciences of their own days; and we should be doing likewise. Therefore, if our science is committed to the existence of entities of some kind, then we should, at least initially, be disposed to accept that they are real. If our science gives us no reason to believe in the existence of things that seem to be required for logical theory, then we should, at least initially, be disposed to deny their reality.

The metaphysical realists of the 1980s, and before, were mostly realists about objects and events, but nominalists concerning properties and relations. For they tended to identify properties with monadic predicates and relations with predicates of higher adicities. But such nominalism is contrary to the spirit of scientific realism, and most realists nowadays hold what David Lewis once called 'sparse' theories of properties and relations. For most would agree that there are predicates in every language of every adicity that do not name real properties and relations, and in all probability there are properties and relations of many adicities that cannot be named in any language.

Many of the metaphysical realists of earlier generations were also first-order extensionalists. That is, they rejected modal languages as 'second-grade' discourse, and denied any significance to quantifications over properties or relations. But such 'strait-jacket' realism is not scientific realism, and the metaphysical theory required for modern science, cannot plausibly be identified with that of the semantic models for certain kinds of logically simple languages. Science needs to be able to refer to natural necessities and possibilities, as well as to causal powers, capacities and propensities. It also needs languages with counterfactual conditionals and objective probabilities, none of which is available in any first-order extensional language.

Metaphysical realists today embrace causal powers and other powerful properties, and are mostly realists about spatio-temporal and numerical relations. They also have views about what these powers and relations are. According to some they are classical Aristotelian universals. According to others, the entities required, especially for mathematics, are Platonic universals. But the movement is not a scholarly one, and there is nothing sacrosanct about these classical concepts. Indeed, it would be surprising if the concepts required for a metaphysical theory designed to accommodate the entities postulated in the sciences and mathematics of antiquity, should also prove to be satisfactory for our modern scientific understanding.

Scientific realism became an important topic in Australasian philosophy many years ago, with the publication of Jack Smart's *Philosophy and Scientific Realism* (1963). For this publication led immediately to a vigorous debate in Australia and elsewhere about the status of mental phenomena. In this book, Smart defended the then seemingly indefensible mind-brain identity thesis, which had been proposed earlier by U. T. Place (1956). Smart's original position developed in his book involved commonsense realism about the substantive theoretical entities of the advanced sciences, which, in itself, would not have caused much of a stir. But his defence of central state materialism made his theory a challenging one. Smart's main argument for scientific realism was an argument to the best explanation. He argued that if things do in fact behave as if theoretical entities of the postulated kinds exist, then the best explanation of this fact is that they really do exist. Therefore, he concluded, if we are to have a rational system of beliefs about what there is in the world, it had better be one that admits physical things such as these. That is, it would have to be a kind of physicalism. So, for Smart, the only question was whether a purely physicalistic ontology would prove to be adequate. Could it, for example, accommodate mental experience? Smart thought it could, and proceeded to spell out his own special brand of scientific realism, in which he identified mental events with brain processes. The philosophy of scientific realism that he developed became known as 'Australian materialism'—or the 'Australian heresy', depending on your point of view.

Interest in Australian materialism was reinforced, but taken in a new direction in David Armstrong's *Universals and Scientific Realism* (1978). Armstrong thought that the arguments for scientific realism were also good arguments for realism about the kinds of properties that scientists needed to postulate to describe the theoretical entities they believed in. For these properties would have to be natural properties, i.e. properties that things had by nature; they could not just be predicates. Predicates, he said, are human inventions—linguistic entities, not fundamental existents. Inevitably, this observation led to a whole new debate about the ontology of scientific realism, e.g. about what kinds of things natural properties really are.

According to Armstrong, natural properties are real things existing in the world independently of human languages, conventions and interests. There are predicates that do not name properties or relations, he thought, and probably many properties or relations that are not yet named. The natural properties, he argued, are of the nature of Aristotelian universals, and so cannot exist uninstantiated, as Platonic universals (i.e. Forms) supposedly can. These natural properties, Armstrong argued, are all categorical, i.e. have identities that depend only on what they are, not on what they dispose their bearers to do. The so-called dispositional properties of things, whose identities do depend on what they dispose their bearers to do, must all supervene on the categorical ones, and on the laws of nature. For, if this were not true, he argued, then the laws of nature would not all be contingent. Following Michael Tooley and

Fred Dretske, Armstrong (1983) argued that the laws of nature must be just contingent relations of natural necessitation between Aristotelian universals. His present view, which he outlines briefly in his paper in this volume (2), is a development of this original one.

Ellis took the debate about the ontology of scientific realism in another direction. Breathing new life into the ancient theory of essentialism, Ellis argued, as Rom Harré and Edward Madden (1975) had before him, that the arguments for scientific realism require that one be a realist about many of the causal powers and other dispositional properties that feature in scientific explanations. For the most fundamental properties in nature, other than the relational ones, would all appear to be dispositional. And, as Ellis and Lierse argued in their joint paper 'Dispositional Essentialism' (1994), the case against dispositional realism is unsound. According to Armstrong, and other categoricalists, all genuine dispositional properties must have categorical bases. But this appears to be an unjustifiable claim, and one that is grossly implausible for fundamental particles. For if these particles are truly fundamental, then there can be no underlying structure to provide this basis. Lierse and Ellis argued that if the basic properties in nature really were dispositional, then the most fundamental things in nature would have to be identified, not by their shape, size or any other categorical properties, but by how their properties dispose their bearers to behave. Moreover, if the most fundamental things in nature are all members of natural kinds, as indeed they would all appear to be, then these things must be bound to behave as they do in the circumstances in which they exist. For they could not behave otherwise without ceasing to be things of the kinds they are. And, if this were the case, then the laws concerning the behaviour of such things would not be contingent, as Armstrong and most others have always assumed, but metaphysically necessary.

At the level at which it makes any sense to be talking about objects with properties, the most basic properties in nature would appear to include all of the causal powers, capacities and propensities of the fundamental particles. It is true that at any more fundamental level than this, we would have to give up thinking about the world as made up of objects with properties, as James Ladyman and Don Ross (2007) have recently argued. But a theory about the nature of reality that would be appropriate for the bizarre world of quantum field theory, which is what we would be left with, would be practically useless for anything else. To a very high degree of approximation to the truth, the real world is one of objects with characteristic properties, behaving just as we should expect things with these properties to behave in the given circumstances. In our view, it is legitimate to consider the metaphysical implications of the form of scientific realism that is adapted to this practical level. Any deeper analysis would lack explanatory power at the relevant level. Any shallower one would be too superficial. The issue of levels of analysis is discussed in Patrick McGivern's paper (4) in Part II of this volume.

Several Australian philosophers have been involved in the project of trying to develop an adequate theory of properties and relations. In the 1990s,
Charles Martin and the late George Molnar were doing pioneering work developing such theories. But much of the work that has been done recently on the
distinction between categorical and dispositional properties has been done as
part of the ongoing work of the British metaphysicians. This volume brings up
to date debates on a range of topics in metaphysical realism, stemming from
the Australian and more recent British traditions.[1] Alexander Bird (2007a) and
Stephen Mumford (2004) have been defending the position that Mumford calls
'pan-dispositionalism', which denies that there are any categorical properties
that are not ultimately reducible to dispositional ones. Bird's paper (3) defends
this position against Armstrong's categoricalism (2), and Ellis's more accommodating theory (1), which accepts that both kinds of natural properties exist.
Ellis argues, as Molnar (2003) did before him, that an ontology for scientific realism needs to include, not only the dispositional properties of the fundamental particles as basic, but also the spatio-temporal and numerical relations that
hold between them.

Ellis's view is that the causal powers in nature all have categorical dimensions, which the laws of action of the causal powers involve essentially. Dimensions, which include all of the natural quantities, are like generic universals. But they are more fundamental than any Aristotelian generic universals.
For they cannot be constituted by their instantiated species, as Aristotelian
generic universals can be. Dimensions have values, he says, which may or
may not be instantiated. Hence, the values of the categorical dimensions of
the causal powers are Platonic universals, not Aristotelian ones.

If Ellis is right about this, and the categorical dimensions are fundamental
to a realist ontology, then we may expect to see important developments on
two fronts in scientific metaphysics—in the field of numerical relations and in
space-time theory. In his book *The Reality of Numbers: A physicalist's philosophy of mathematics*, John Bigelow (1988) has already broken new ground in
developing the required kind of theory of numerical relations. Stathis Psillos
(5), and Anne Newstead and James Franklin (6), both develop theories about
how adequate metaphysical foundations of mathematics may be generated.
Psillos argues that what is required is a Platonistic metaphysics of scientific
realism. For the only alternative, he says, is a nominalistic one, which is unsatisfactory. Newstead and Franklin are likewise anti-nominalistic, but they seek
to break the back of the argument for Platonic realism by adopting Armstrong's
criterion for metaphysical realism, viz. that 'to be is to be a truthmaker (or part

[1]This collection derives from a conference held at the University of Melbourne in July 2009 to
discuss issues relating to the metaphysics of science. This conference was sponsored by La Trobe,
Monash and Melbourne Universities, and the Centre for Time in Sydney. The papers included
here were written especially for the conference, and all of them presuppose scientific realism.
The issues discussed in this volume are those concerning the implications of scientific realism for
metaphysics. The conference was partially co-extensive with the annual workshop of the British
Metaphysics of Science project, supported by the U.K.'s Arts and Humanities Research Council.

of one)' to replace W. V. O. Quine's 'indispensibility' requirement, viz. that 'to be is to be the value of a variable', and developing a neo-Aristotelian form of realism, which, they argue, is itself adequate for mathematics.

Stephen Mumford and Rani Lill Anjum (7) develop the theory of causal dispositionalism. Theirs is the first of three papers on dispositions and causal powers included in this volume. The others are those of Max Kistler (8) and William Bauer (9). Mumford and Anjum argue that dispositional properties, or 'powers' as they are sometimes called, are basic in ontology, and adequate to ground a theory of causation. Their aim of developing a theory of causation from an ontology of powers was anticipated by Rom Harré and E. H. Madden (1975), and shared by Molnar (2003: 186) and many others. But Mumford and Anjum claim that the 'delivery of a plausible powers-based theory of causation is ... overdue. So far', they say, 'we have only hints and false starts. None of these accounts has gone quite in the right direction.' Their aim in this paper is to redirect the inquiry.

Max Kistler's paper (8) is specifically concerned with the theory of dispositions and their grounds. Kistler argues that there is an important difference between the dispositions of things and the powerful properties that are needed to explain them—one that is often overlooked. The latter are plausibly real, whether or not they have what some would call a categorical basis, or are explicable by reference to more fundamental properties. The former, i.e. the dispositions, are not plausibly real properties, but just observed patterns or regularities. The importance of the distinction lies in this fact. For, if Kistler is right, then the commonly defended positions of dispositional realism and reductionism about dispositions would appear to be compatible. Realism is right, one could say, about powerful properties, reductionism is right about dispositions.

William Bauer's paper (9) discusses four contending theories of pure dispositions, i.e. dispositional properties that require no causal basis, either in other dispositions or in categorical properties. The thesis that such properties exist faces the Problem of Being: Without a causal basis, what ontologically grounds the continued existence of a pure disposition when it is not manifesting? After explaining the Problem of Being, this paper argues for a distinction between the causal basis and the ontological grounds of a disposition, suggesting that a pure disposition may be grounded yet remain pure because it does not have a causal basis for manifesting. Given this distinction, the paper evaluates four theories of the grounding of pure dispositions: (i) that pure dispositions are grounded globally in all properties; (ii) that pure dispositions are grounded in the world as a whole that is ontologically prior to the world's parts; (iii) that pure dispositions are grounded by their object-bearers; and (iv) that pure dispositions are self-grounded properties. The paper argues that (iv) is the most viable, and develops an explanation of self-grounding. The core principle is that a pure disposition grounds itself via a minimally sufficient occurrence of its own power, as opposed to one of its characteristic manifestations.

The section that follows is concerned with the theory of pan-dispositionalism. Three papers on this topic are included here, viz. the papers by Matthew Tugby (10), Sharon Ford (11) and Barbara Vetter (12). Tugby's paper is concerned with the theory of 'pan-dispositionalism'. He distinguishes two versions of it, one involving universals, and the other sets of tropes, and argues that a pan-dispositionalist has special reasons for viewing dispositional properties as universals. He says that they should specifically be seen as universals that are at least partly constituted relationally, and argues for this on the ground that, with universals in play, pan-dispositionalists can successfully give an account of the directedness of dispositions, and do so despite the fact that instances of dispositions are often intrinsic to their possessors, and may exist even if they are never manifested. In contrast, it is not clear that a 'trope' pan-dispositionalist can account for these facts satisfactorily. Tugby concludes his paper by focusing in more detail on the 'universals' version of pan-dispositionalism and considering the nature of the relations that, on this picture, hold between property universals. He suggests that these second-order relations must be internal in some sense, depending on which version of pan-dispositionalism is adopted.

Sharon Ford's paper (11) is also concerned with pan-dispositionalism, but her concern is with the viability of a theory of reality that attempts to construe all spatiotemporal relations dispositionally. Several writers (e.g. Mumford 2004; Bird 2007) have noted that the laws of nature do not seem to make any distinction between categorical and dispositional properties. The gravitational masses of things, and their spatial separations, both seem to contribute essentially to Newton's law of gravity, and apparently do so in much the same sort of way. If there are effects of the masses on the strength of the attractions between bodies, then there are also effects of the distances separating them. So, it is at least plausible to argue that distances between things are no less causally efficacious than gravitational masses. This is not quite the same thing as arguing for pan-dispositionalism, since the argument to this point is purely formal. But if no satisfactory theoretical basis for a distinction between categorical and dispositional properties can be found in science, this would seem to be a good reason to suppose that none exists.

Barbara Vetter's paper (12) looks at the argument for Dispositional Essentialism (DE) that has been put forward by Bird (2007a) in his recent book *Nature's Metaphysics*. Bird's overall argument comes in two parts, one negative and one positive, which together are supposed to establish DE as the best contender for a theory of properties and laws. Vetter argues that, even if all their particular steps go through, both parts of the argument have significant gaps. The negative argument, if successful, shows that at least one property has an essence, but not that every property has a dispositional essence, as Bird maintains, and as every pan-dispositionalist believes. The positive argument, which aims to demonstrate the explanatory power of DE, fails to take account of the quantitative nature of the fundamental natural properties and laws. Vetter concludes her paper by suggesting a revision of DE that might solve this problem.

Finally, there are two important papers concerned with the theory of natural kinds. In his *Scientific Essentialism*, Ellis (2001) argued that the natural kinds in any given category form a hierarchy, ranging from a global kind, which includes all natural kinds of things in this category, down (by the species relation) to a set of infimic species (i.e. species that have no sub-species) of this global kind. Armstrong, on the other hand, argued that all generic or determinable kinds supervene on the simple monadic and relational universals that are these infimic species. These theories of the relationships between generic and specific natural kinds are challenged in these two papers. In her paper, 'The Metaphysics of Determinable Kinds', Emma Tobin (13) argues, against Ellis, that there are many theoretically important natural kinds that do not fit into his neat hierarchical structure, including, among others, the functional kinds of the biological and social sciences. Ellis and Armstrong agree that our scientific worldview should inform our general ontology, but Tobin argues that a more complex view of the natural kinds is required than either Ellis or Armstrong has provided.

Corinne Bloch (14) argues that a metaphysics of scientific realism need not be an essentialist one. For one can accept a scientific world-view, and provide adequate accounts of scientific kinds and their definitions without assuming that any of these kinds of things have mind-independent intrinsic essences, as Ellis requires of natural kinds. To establish her point she uses case-studies from 20th century neurobiology to argue that many scientific kinds are not natural kinds (as Ellis defines them). In the concluding sections of the paper, she argues that these results can be generalised.

While many of the issues under debate have ancient roots, these essays fully demonstrate that metaphysics need not fear naturalism. On the contrary, metaphysics is amply able to respond to the demands of science and with modern science to articulate a coherent world-view. The details of such a view are areas of exciting current research and scholarly dispute, but the project of a realistic metaphysics of science is surely well advanced.

Part I

SYMPOSIUM ON PROPERTIES

THE CATEGORICAL DIMENSIONS OF THE CAUSAL POWERS

Brian Ellis

The causal powers of things are all dispositional properties, i.e. properties whose identities depend on what they dispose their bearers to do. But I do not think, as some philosophers do, that all properties are dispositional. Spatio-temporal relations, for example, may hold, without disposing their relata to do anything. Nevertheless, there is an important link between these relations and the causal powers that things may have, because the instances of the causal powers must all have categorical dimensions, i.e. spatio-temporal locations, distributions, and so on, which are definable only in terms of such relations. Formally, the structures definable wholly in terms of such relations are just quiddities, i.e. things whose identities depend on what they are—not on what they do. But this is not to say that they have no place in a realistic theory of what there is. For, if there were no quiddities in the world, there would be no causal powers. And, if the there were no causal powers, then the world itself would be just a quiddity.

1 Physical causal processes

For a causal realist, the class of causal powers must be defined with reference to that of physical causal processes, since, whatever else they may be, the causal powers of things must at least dispose them to take part in such processes. Therefore, if we wish to define the class of causal powers, we had better begin by attempting to define this class of physical processes. To define it, it is important to distinguish between what a thing causes, and what a person may be held responsible for. Responsibility is a moral or intentional concept, not a purely physical one. So, it is not a concept of the required kind for causal realism. Physical causation may (or may not) be involved when we are doing what we are intending to do, or what we are morally responsible for doing. But, in any case, the intended outcomes, or those for which we are morally responsible, may well be due directly to the actions of others, and so not physically caused by us. To find genuine examples of physical causation we must look at what happens in the inanimate world, where there are no intended consequences. In this world, the mechanism for causation appears to be simply that of energy transfer. The system that acts, i.e. is doing the causing, transfers energy to the system that is acted upon, i.e. is being affected. Thus, in Hume's billiard ball example some or all of the kinetic energy of the impacting ball is transferred to that of the ball impacted upon. In the act of warming oneself in front of the fire, the chemical energy stored in the wood is released as heat in the process of burning, and the heat energy is transferred radiantly to the person who is warmed. Such examples are obviously causal processes, and the mechanism of causation is no less obviously that of energy transfer. Therefore, it is initially plausible to suppose that this is the physical mechanism that we are seeking.

It is true that not all physical causal connections seem to be quite like this. In many cases, for example, the causal process is two-way. Thus, electrons repel each other, and gravitational masses attract each other. But, in these cases, it is now widely believed that there is an exchange of virtual particles—photons in the first of these two cases, and gravitons in the second. So, these cases do not seriously undermine the initially plausible suggestion that physical causation is essentially an energy transfer process. A more serious threat to the energy transfer theory derives from interventions. For we all want to say such things as: the eclipse of the sun caused the light to fade, or the dark glasses filtered out the ultraviolet light. And these would all appear to be physical causal notions. But dimming, shading, filtering out, deflecting, and so on, are all negative causal influences. They affect the system that is influenced by preventing an energy transfer process from occurring or having its full effect.

Let us, therefore, distinguish between positively and negatively acting causes, and recognise that most of the processes we call 'causal' involve elements of both. But let us not be too distracted by this. Our natural tendency is to focus on the actions that we may take (e.g. shading our eyes), and the consequences that these actions may have for us (e.g. reducing the glare). For these are what interest us. But the underlying physical causal process remains the same. It begins with radiation from a light source, and it ends on whatever absorbs this light. And this remains the case, whatever I may do to shade my eyes. Therefore, the direction of energy transfer remains the same, even though the amount, or nature of the energy transferred is reduced or changed. So, I am going to bite the bullet on this, and say that shading X is not the direct cause of anything that happens on X. It may be preventing something from happening on X. But preventing is not a special case of physically causing. Nor is deflecting, filtering out, or any of the other ways of interfering with physical causal processes. The so-called 'negatively acting causes' are negative in conception, but the underlying directions of the energy transfer processes involved, and hence of the physical causal influences they have, remain the same. Moreover, the negatively acting causes achieve their effects in very different ways from the positively acting ones. If I shade my eyes, I prevent some or all of the light from reaching them. But this result is effected by doing something positive, viz. absorbing or reflecting some or all of the light coming towards my eyes. It is not achieved by doing anything to my eyes.

Accordingly, we may define a physical causal process A$\Rightarrow$B as an energy transference from one physical system S_1 in a state A to another physical system S_2 to effect a physical change B in that system (relative to what the state of S_2 would have been in the absence of this influence). Note that, given this definition, the direction of causation is necessarily that of the energy transfer. Negative causal influences notwithstanding, if there is no energy transfer from A to B, then there is no physical causation in that direction. In the simplest kind of case, an energy transmission from one thing to another consists of a particle emission (e.g. a photon), its transmission in the form of a Schrödinger wave (e.g. an electromagnetic wave), and its subsequent absorption (e.g. on a photographic plate, or any other particle-absorbing surface). I call all such processes 'elementary' causal ones, and postulate that all physical causal processes consist ultimately of just such elementary processes as these. We know that these elementary causal processes are all temporally irreversible, because the process of Schrödinger wave collapse is temporally irreversible. For there is no such thing as the instantaneous, or near instantaneous, reflation of a Schrödinger wave. Therefore, for a realist about Schrödinger waves, it is reasonable to suppose that all physical causal processes are temporally irreversible, and that the temporal order is the same as the physical causal order.

The level of analysis required for the description of Schrödinger waves is at what we may call 'the base level'. It is not, however, at the level that is most appropriate for the description of physical causal processes. It may be taken as establishing the temporal asymmetry and temporal direction of physical causation. But it is much too deep for the purposes of this paper. At this level, even the physical objects that are the bearers of causal powers do not exist. However, physical objects capable of having causal powers manifestly do exist at what I shall call 'the object level', i.e. at the level of ontological reduction at which we may speak freely of physical objects and their properties. The question of how things at this level may ultimately be constituted by elementary events and processes will be set to one side for the time being—although I shall have something to say about this question towards the end of the paper.

The physical causal processes that are the displays of causal powers are presumably ones that belong to natural kinds. For, if they were not members of natural kinds, the causal powers would not be natural properties. Nevertheless, there are clearly a great many natural kinds of causal processes in nature, and so, presumably, a great many natural causal powers. All of the chemical reactions described in chemistry books, for example, are processes that belong to natural kinds; and the equations that describe these reactions are presumably descriptive of their essential natures. There are also a great many natural kinds of physical causal processes occurring in other areas of science. So there is no shortage of causal powers.

2   Causal powers

At the object level, there are apparently two very different kinds of properties, dispositional and categorical. The dispositional properties are those whose identities depend on what they dispose their bearers to do, and the categorical ones are those whose identities depend on what they are—but not, apparently, on what they do. The latter are, for the most part, the spatiotemporal and numerical relationships that are required to describe the structure of things. The causal powers, on the other hand, belong in the category of dispositional properties, along with propensities (See Section 8 below), which, I shall argue, are not causal powers. What these active properties all have in common is their dispositionality. The categorical properties, in contrast, are essentially passive; since there is nothing that their bearers are necessarily disposed to do just in virtue of their having these properties. Nevertheless, I will argue, the categorical properties do have some vital causal roles. For these properties determine where the active properties of things may exist, or be distributed, and, consequently, where the effects of their activities can be felt.

Given this preamble, we may define a causal power as any quantitative property P that disposes its bearer S in certain circumstances C_0 to participate in a physical causal process $\Rightarrow$, which has the effect $E - E_0$ in the circumstances C_0, where E is the actual outcome and E_0 is what the outcome would have been if P had not been operating. In general, the changes that P induces/prevents in the circumstances C_0 will depend on the measure of P. Let P_0 be the measure of P for S in the circumstances C_0. Then, since P is a causal power, P must have a law of action A that describes what it does when it acts. If there were no such law of action, there would be nothing to connect different exercises of the same causal power. Since, by hypothesis, the value of the causal power involved in transforming E_0 into E is P_0, the law of action $A(P, x)$ for P must be one that has the effect $E - E_0$, where $P = P_0$, and $x = C_0$. That is,

$$E - E_0 = A(P_0, C_0).$$

This effect, it will be noted, is a function of two variables, P_0, which is the measure of the causal power in the circumstances, and C_0, which is the set of values of the relevant parameters describing the circumstances in which P is operating. (N.B. E_0, which is the set of circumstances that would have obtained if P had not been operating, would often, but not always, be the same as C_0. But in cases in which the cause would act to enhance or inhibit some effect that would otherwise occur naturally, E_0 and C_0 would normally be different.)

To illustrate: a weight above ground level has causal powers just in virtue of the potential energy it possesses, e.g. it has the power to compress things and the power to stretch them or pull them down. How this power affects other things depends on where it is, and how it is fixed in its position, e.g. on whether it is resting on something, or hanging from it. Suppose the weight is of magnitude W, and that this weight is hanging at the end of a wire S made of some elastic material M. An elastic material is, by definition, one that is, within the limits of elastic distortion, intrinsically disposed to resist distortion, and, if distorted, to revert back to its original shape. If a piece of such material is stretched, for example, then it will be disposed to resist being stretched, and to regain its original shape, once the stretching force is removed. Let S be of length l_0 before the weight is hung upon it. The action of hanging W from S then causes it to become extended, and (provided that the elastic limit is not exceeded) to acquire sufficient potential energy to return to its original length and cross-sectional area once the weight is removed. Let l be the extended length of S. Then the weight loses potential energy $(l - l_0)$ W, while the wire gains exactly this much elastic energy in the process. So, the process is a simple physical causal process, involving a direct energy transfer from the weight to the wire. The causal powers involved are the power to act (gravity) and the power to resist (elasticity).

The law of action for W on S depends on the material of S. Empirically, it is found that the extension of the wire is proportional to l_0 and to W, and is inversely proportional to the cross-sectional area A of the wire. Let P_0 be the elasticity of the material of S. Then, by definition:

$$P_0 = \frac{W}{A} \times \frac{l_0}{l - l_0}$$

Accordingly, the law of action of the causal power (elasticity) must be:

$$(l - l_0) = \frac{W l_0}{A P_0}$$

where $(l - l_0)$ gauges the effect (i.e. the stretching of the wire), W measures the strength of the causal power that is acting, and l_0/A is the relevant function of the categorical dimensions of the initial circumstances.

Like the laws of action of all causal powers, this one is quantitative, depends on the magnitude and location of the powers concerned, and involves one or more other categorical properties essentially. In this case, these other categorical properties are the original length l_0 of S, the amount $(l - l_0)$ by which S is extended, and the cross-sectional area A of S.

3   Dimensions

The dimensions, as I shall define them here, include all of the important quantitative properties involved in the laws of nature, and most, if not all, of the important categorical properties as well. Such dimensions are important, it will be argued, for two reasons: (1) they are the generic properties, sometimes called 'determinables', that are directly involved in the laws of nature, and (2) they are ontologically more fundamental than any of their actual or possible 'determinates', which are what I call their 'values'. But first, some definitions: Let P be any generic natural kind or property. Then,

(1) An *infimic species* of P is any species of P that has no sub-species.

(2) Two things are *specifically the same in respect of* P, if and only if the values of P instantiated in these things are members of the same infimic species.

(3) x is an actual value of P, if and only if x is a value of P that is instantiated.

(4) P is a dimension if and only if there are at least two values of P that are specifically different, and at least one of these values is actual.

The dimensions include all of the usual spatial and temporal relations. For these relations are all quantitative, and all quantitative properties are dimensions. The dimensions also include the generic categorical properties of things, including shape, size, orientation, handedness, spatio-temporal interval, and so on. But they are not restricted to these categories, since the causal powers are dimensional too. They are like ordinary universals in some respects, since they have essences. But they are more general than ordinary universals, since their infimic species need not all be specifically the same. Conceptually, they most closely resemble what I have elsewhere called 'generic universals'. There are, however, several good reasons for breaking with tradition here, and giving them a more prominent role in metaphysics than generic universals have ever had. For dimensions are ontologically more fundamental than traditional universals. Firstly, the specific universals, which are the actual values of the dimensions, could not exist if the dimensions did not exist. Secondly, the dimensions cannot, in general, be constituted by their actual values.

To illustrate: consider the generic property of having mass, and the specific property of having a mass of two grams. The first is an example of what I call 'a dimension', and the second is what I am calling 'a value of this dimension'. Given that something actually has such a mass, this value would be a classical universal. But, fairly clearly, this specific value could not exist if the dimension did not exist. On the other hand, the dimension could exist, even if this specific value did not. Therefore, by the usual criteria for ontological priority, the dimension must be considered to be ontologically prior to this universal, or to any other, of its actual values. Moreover, it is evident that the dimension of mass cannot be constituted by the disjunction of all of its actual values. For, (1) as Armstrong has convincingly argued, there are no disjunctive universals, and (2) not every value of the dimension of mass is instantiated. A quantitative universal constructed disjunctively out of classical (Aristotelian) universals would therefore be likely to be discontinuous, i.e. to have unwelcome gaps in its range.

The main reason for focussing here on dimensions and their values, rather than on universals and their genera, is that the dimensions of things must be seen being as among the fundamental constituents of reality at the object level. So, any modern metaphysics that is designed for talking about the structure of reality at this level should use this language, and think of the properties that are most relevant to this structure as being the dimensions of things, not the specific universals that are their actual values. The laws of nature, for example, are not even plausibly just relations between traditional universals, as Dretsky, Tooley and Armstrong have all maintained. Such relations are too specific to be counted as laws of nature. They are what scientists would call mere *instances* of these laws.

The laws of nature themselves all express generic relations between the quantitative properties of things. And the quantitative properties that are generically related are all dimensions. It is true that if we knew the metaphysically necessary relations between the dimensions of things, then we could easily derive all of the possible relations that must hold between their values. But the laws of nature cannot be constituted by the specific relations that must hold between the actual values of the dimensions, for precisely the reason that these dimensions cannot be constituted by their actual values. So, my first thesis is this: The laws of nature, including all of the laws of action of the causal powers, describe the relations that must hold between the basic dimensions of things.

4   The categorical dimensions of the causal powers

In response to two of my critics (John Heil and Alexander Bird) at the *Ratio* Conference of 2004 on Metaphysics in Science, I argued (Ellis 2005b: 95–6) that,

> ...a property can have a causal role without either being a causal power, or being ultimately reducible to causal powers. For even the most fundamental causal powers in nature have dimensions. They may be located or distributed in space and time, be one or many in number, be scalar, vector or tensor, alternate, propagate with the speed of light, radiate their effects uniformly, and so on. But these dimensions of the powers are not themselves causal powers. A location in space and time is not itself located in space and time. Nor does having a magnitude have a magnitude. Nor is

being one or many in number itself one or many in number. Yet these dimensions of the powers clearly do have causal roles. They not only signify the respects in which causal powers may be similar or different from one another, their detailed specification is required to define the laws of distribution, action and effect of the powers. These dimensions of the powers are the properties that I call categorical. They are real, and no less important in the overall scheme of things than the causal powers that have them essentially. In reality, they are second-order properties—properties of properties. They are indeed amongst the essential properties of the causal powers.

I now disagree with some of this. I was right in thinking that there are categorical dimensions implicit in the causal powers. But I no longer think of the categorical dimensions as second-order properties, as I did then. Locations, which are what I had in mind when speaking of the categorical properties, are (if they are properties at all) properties of *instances* of the causal powers, not properties of the powers themselves. Therefore, they are not second-order properties. On the contrary, they must be ontologically more fundamental than the causal powers. For, without them, the causal powers could have no instances, and so could not exist. Moreover, the instances of the causal powers normally have magnitudes and directions, and usually they are capable of acting together to produce effects that none could produce alone. But without the categorical properties to locate, indentify and orient them in space or time, there could be no laws of directionality, distribution, of combination of the causal powers. They would be nowhere, nowhen, directionless, and lacking identity.

5   The Importance of location

Every instance of a causal power must have some specific location. It must be a causal power of one thing rather than another, or be located in one place rather than another, or be distributed in this region rather than that one. But no instance of location has any specific location contingently. It is just where it is—and necessarily so. A specific instance of any genuine causal power, such as gravitational mass, could be anywhere. Its whereabouts is contingent, and it might have any of infinitely many possible locations. But no specific instance of location has its location contingently. Necessarily, it is where it is. Therefore, location cannot possibly be a genuine causal power. Moreover, if you remove all of the causal powers from any given location (and this would always seem to be a possibility), the location remains, but the causal powers do not. Therefore, it cannot be said truly that the locations of things have any causal powers essentially.

But if location is not a causal power, and locations do not have any causal powers essentially, then what are they? According to Heil (2005) and Bird (2005a), they must be quiddities, and therefore unreal. And, presumably, any properties or structures that are definable only in the sorts of ways that locations are definable, must also be quiddities.

However, I am not at all disturbed that some important properties should be just quiddities. In fact, I think that categorical properties and structures are much more like locations than like causal powers. For they are all quiddities. At least, this is my hypothesis. It is a plausible hypothesis, because none of the properties or structures that are definable in terms of spatio-temporal relations would appear to be causal powers. On the contrary, the shapes, sizes, orientations, motions and so on that are fully definable spatio-temporally all seem to lack the essential properties of such powers. None of them has a plausible law of action, or produces any effect that might be due to its action. Nor are the categorical properties of things powers to resist, deflect or otherwise interfere with the actions of any known causal powers, as, for example, elasticity, inertial mass, and electrical resistance clearly are. They would all appear to be just categorical properties, i.e. properties whose identities depend not on what they do, but on what they are. So, my hypothesis is that the categorical and other structural properties of things, which are definable spatio-temporally, are all quiddities, just as Heil and Bird say they would be, if they existed, and were not reducible to, or otherwise able to be construed as, causal powers. I agree with them about all this. But, unlike them, I think that categorical properties and structures really do exist, and that they are all quiddities.

6   Against quidditism

The fundamentally Lockean position outlined here has been attacked on two fronts. There are strong categoricalists, who deny that there are any genuine causal powers, and strong dispositionalists, who deny that there are any real categorical properties. I have discussed the principal arguments for and against strong categoricalism elsewhere, and I shall not repeat this discussion here. The main objection to my position now appears to be coming from strong dispositionalists, who argue that (a) if the categorical properties of things had no causal powers, then we should not know anything about them, and (b) if the causal powers of the different categorical properties were not distinctive of them, then we should have no way of distinguishing between them. Therefore, they say, the categorical properties of things must all have or be distinctive causal powers, i.e. powers that would distinguish them essentially from one another. Otherwise, it is said, they would all be mere quiddities.

This argument from quidditism against the possibility that we could have any knowledge of categorical properties seems, at first, to be very persuasive. But we should not be too readily persuaded by it. For, the categorical properties of things can always be effective through the laws of action of the causal powers, even though they are not themselves causal powers. The magnification achieved by a lens, for example, is a function not only of the focal length of the lens, and hence of its of magnifying power, but also of its distance from the object that is being magnified. So distances can be factors in determining outcomes, even though the distances in question are not causal powers. One possibility, therefore, is that categorical properties are much more like the locations of instances of the causal powers than they are like the causal powers themselves. This is the position I wish to defend.

But have I not, in stating my position, already conceded its falsity? In allowing that distances can be factors in determining outcomes, have I not already conceded that distances can have causal powers? I deny this. Being a factor in determining an outcome is not the same as being a cause, or having the causal power to achieve this outcome—at least, not as I have defined physical causation. Living a long way from Sydney is a factor in determining whether or not I can walk there. But it is not a physical causal power, or anything that has such a power. It is also a factor in determining whether or not I can get there by road in less than ten minutes. But it is not a cause of my not being able to get there by road in less than ten minutes. Lengths, orientations, distances, times, shapes, and sizes may all be factors in determining what outcomes are certain, possible or impossible. But this does not imply that they are causal powers, or that they have causal powers.

Most of the properties we think of as categorical are ones that depend upon our being able to recognise common patterns of spatio-temporal relations. It is not my firm view that they are all dependent upon such patterns of relations. For I have no proof that this is so. But certainly a great many of them are: e.g. shape, size, orientation, speed, handedness, direction, angular separation, and so on. And, like the spatio-temporal relations upon which they depend, they are fairly clearly not causal powers. It could perhaps be argued that spatio-temporal relations are not causal powers because they are not properties. Hence, the fact that the dimension of spatio-temporality is not a causal dimension might easily be reconciled with a strong form of dispositionalism. All genuine properties, it might be said, are causal powers.

However, the properties I am calling 'categorical' are not out-of-the-way, recherché properties. Every law of action of every causal power refers to some of the categorical properties of the circumstances in which it may be effective, and the properties that I have listed above are among those that are most often required to describe these circumstances. So anyone who uses the 'not really properties' defence would owe us an account of what the categorical properties really are. And, to their credit, I do not know of any strong dispositionalist who has ever used this argument. The argument on which they seem to rely most heavily is the one from quidditism, which, they think, settles the matter.

Thus, in his defence of Sydney Shoemaker's ((1980)) strong version of dispositional essentialism, Bird (2005a) argued that what makes a property the property it is, i.e. what determines its identity, is its potential for contributing to the causal powers of the things that have it. Bird's case for this strong version was that anything weaker would condemn us to quidditism. For if there were any property whose identity did not depend on what it disposed its bearer to do, but only on what it was (quidditism), then we should, necessarily, be ignorant of it. So, Bird is evidently committed to arguing either (a) that location is a causal power, or (b) that location has at least some causal powers essentially. But, as we have seen, location is not a causal power, and locations do not have any causal powers essentially. Therefore, locations are all quiddities.

But if locations are all quiddities, then so are relative locations. For the actual locations of things depend essentially on their locations relative to things whose actual locations are taken as given. That is how they are defined. And, if these relative locations had causal powers, then we could reasonably argue that the actual locations of things must also have causal powers. But we have already excluded this possibility. Therefore, relative locations neither are nor have any causal powers. Therefore, the shapes of things, which depend essentially on the relative locations of their parts, must all be quiddities too. And, if this is so, then, plausibly, the same must be true of all of the categorical properties. They must all be quiddities. Does this mean, then, that we cannot know anything at all about them? I think not.

7   In favour of quidditism

It is commonly thought that if the world contained any quiddities, then they would necessarily be unknown to us, and hence that any claim to knowledge of them would have to be false. But this is not what follows from categorical quidditism. It just means that that the shapes, sizes, orientations etc. of things cannot be known without the mediation of the causal powers that are located within them. But since the objects that we can or do know about do have causal powers, the quidditism of their categorical properties does not matter. In fact, the categorical properties have long been recognised by empiricists as being among the most directly observable of all of the kinds of properties.

But, how could anything that does not have any effects essentially be observable? My answer is straightforward. Accidentally. It could, like location, do so by determining where the causal powers act from, and so where their effects may be felt. Or it might determine how strongly, or in what manner its effects may be felt. The shape of an object itself has no effects essentially (since it does not itself do anything), but it does partly determine the shape of the pattern of effects accidentally produced by the reflective powers of its surface material. For the shape of the object is one of the factors determining the spatial distribution of these powers. An unilluminated square has no visible effects. But an illuminated square does have such effects. It looks square. It is true that it does not have this effect essentially. For it is not essential to the squareness of an object that it should look square, or even that we should be capable of vision. The squareness of the object is not a source of the energy transfer processes that result in this perception. The lights that enable us to see, and the reflectivity of the various parts of the object's surface, are the sources of these. The shape only determines where the reflected light may come from. But, like location, the object's shape is inert. It is just a quiddity.

The causal powers of the lenses of our eyes produce images on our retinas of the shapes of things in our fields of vision, and the causal powers of the rod and cone cells in our retinas reproduce encoded versions of these retinal images in our occipital cortices, which, presumably, then ramify through the integrative circuits of our brains to produce conscious awareness of the shapes that we are observing. So, there is no great mystery about how we are able to know about the categorical properties of things in our environments. There would only be a mystery if there were no sources of power to illuminate the objects we see, or to reflect the light from their illumination, or if we ourselves lacked the visual or mental capacities to pass on or process the encoded information that we receive when we look at things.

There is no need, therefore, to be afraid of the categorical quiddities of nature. On the contrary, they are among the most direct objects of knowledge that we have of the world. They are not the only such objects. For the causal powers of things also give us direct knowledge of the world. They colour the world that we see, provide taste to the world that we savour, material presence to the world that we feel, and so on. But this direct knowledge is much more ambiguous. The colours are normally mixtures of many different kinds of light, and the tastes are usually our overall responses to many different kinds of substances. There is no good reason, therefore, to think that all properties are causal powers. Indeed, the best explanation that we have of the content of our sense experience derives from the view that there are two kinds of properties, categorical ones, as well as causal powers.

Some of those who wish to say that all of the most basic properties in nature are causal powers may have another reason, besides fear of quidditism, for holding their position. This is the worry that if any of the basic properties in the world are categorical, then many of the fundamental laws of nature, viz. those involving these properties, must be contingent—contrary to the theory of scientific essentialism. For me, this would be a worry, if it had this implication. If the most fundamental of the substantive natural kinds had only structural, and therefore categorical, properties essentially, the substantive natural kinds would not have any causal powers essentially. Therefore, any laws of nature involving such entities would have to be contingent, contrary to the theory of scientific essentialism. However, this worry is unfounded. For the evidence is that the most fundamental of the substantive natural kinds do have causal powers essentially—properties such as mass, charge, spin, and so on, which are essentially dispositional. So there is no good reason to think that more complex objects, which consist essentially of structures of more elementary ones, do not have their causal powers essentially.

It is true that complex chemical substances are normally identified structurally. When a biochemist makes a new molecule in the laboratory, she is likely to have a fairly good idea of what its molecular structure is. For she is bound to have a good understanding of the molecular structures of the substances she has used in its manufacture, and of the likely chemical consequences of the procedures she has employed. If confirmation is needed, the techniques of X-ray crystallography may be used to build up an accurate picture of the new molecule she has made. It is tempting to say, therefore, that the essence of the new molecule depends only on the spatial relations between its parts. But the identity of a molecule cannot depend only the spatial arrangements of its unidentified atoms. It must also depend on what kinds of atoms are involved in this structure, how they are arranged, and how they are bound together. For, until we know all of these things, we cannot deduce anything from a knowledge of the categorical structure alone about how it would be disposed to behave in laboratory tests. We need not, therefore, be disturbed by the thought that the new molecule will have no causal powers essentially. Its causal powers will have to be discovered, of course, and explained, just as other metaphysical necessities must be. But the empirical nature of the investigation that is required does not impugn their necessity.

Suppose that the new molecule S_0 is found to have the causal power P, which is characterised by the law of action L, according to which S_0 will react with substance S_1 having the molecular structure $M(S_1)$ to produce a substance S_2 with the molecular structure $M(S_2)$. Then the law of action L will be metaphysically necessary, and hence the transformation by S_0 of the structure $M(S_1)$ into the structure $M(S_2)$ will also be metaphysically necessary. The fact that L has categorical dimensions is therefore irrelevant to its metaphysical necessity.

8 Propensities and monistic dispositional essentialism

Propensities are dispositional properties, since their identities depend on what they dispose their bearers to do. But, unlike the causal powers, they are unconditional, and the activities of their bearers do not depend on the circumstances of their existence. They are what might be called 'unconditional' dispositional properties. Some examples are: half-lives, radio-active decay potentials, excitation levels, and the particle realisation potentials of quantum mechanics. These properties are certainly dispositional, but they are not causal powers, as causal powers are here defined. For the processes of radioactive decay photo-emission and Schrödinger wave absorption are not causal processes. A substance that undergoes radioactive decay does so independently of the circumstances of its existence, and there is no describable energy transfer process that connects the radioactive particle to its decay products. At the moment of radioactive decay, an instantaneous change of state occurs, in which the decay products come immediately into being. But there are no temporally extended processes by which this change of state comes about. The gamma rays (if any) that are emitted, do not take time to get up to speed. Nor do the decay products take time to form or be ejected once the process begins. The same is true of photo-emissions. The excited atom emits a photon and falls to a lower level of excitation. But there is no causal process by which this change of state comes about. It just happens.

Schrödinger wave absorption is another process of this kind. There is a certain probability of an absorption event occurring—defined by the absolute value of the wave amplitude at the relevant point. But the change of state that occurs when a Schrödinger wave is absorbed is instantaneous, and the point of absorption is localised. But, as in the radio-active decay processes, there is no temporally extended process by which this change of state comes about. It just happens. According to quantum mechanical theory, the mechanism of energy transfer between systems, which is of the essence of causality, is the Schrödinger wave. Minimally, a causal process involves an emission event, an energy transference, and an absorption event. But the emission an absorption events are not themselves causal processes. They are among the *ingredients* of such processes. Hence, the emission and absorption potentials of quantum mechanics are not causal powers, as causal powers are here understood, but something more primitive than causal powers.

In Section 2 above, it was argued that there cannot be a viable form of strong dispositionalism that is based on an ontology of primitive causal powers. For all instances of such powers must be located, and their laws of action must depend on the categorical properties of the circumstances in which they would be active. But propensities are more primitive than causal powers, and their laws of action are independent of circumstances. Therefore, if there is a viable form of strong dispositionalism it must be one that is based somehow on an ontology of propensities. I do not see how to construct such an ontology. For I do not see how to explain what the categorical properties might be, given such a basis. But, I cannot rule out the possibility of such a thesis one day being developed.

Meanwhile, I think it is sensible to accept, at least provisionally, a less ambitious ontology that includes both causal powers and categorical properties. Strong categoricalists, such as Armstrong, imagine that they live in a world of quiddities. Their main problem is to explain what causal powers really are. For they certainly need them. Strong dispositionalists, such as Bird, imagine that they live in a world of causal powers. Their main problem is to explain how causal powers can be located, and have effects that are dependent on their locations. For they certainly need to capture locations in their ontologies. But for causal realists, like me, who distinguish between categorical properties and causal powers, it is relatively easy to give a satisfactory account of both, and have the best of both worlds.

9   Categorical structures and causal powers—an overview

I have assumed, so far, that categorical structures and causal powers are members of the same ontological family. But, if they are as different from one another as I have presented them as being, the question must arise whether they really do belong together. Although I reject both Armstrong's and Bird's theories of properties, I am sympathetic to their attempts to construct a general theory. My own theory is not a general one, and, as it stands, is metaphysically unsatisfactory. Let me therefore toy with the idea that causal powers and categorical structures are members of different categories of being. We know, to begin with, that they are essentially very different from one another. The structures are quiddities, and so are essentially passive, whereas the causal powers are dispositional, and essentially active in some circumstances. It is prima facie implausible, therefore, that there could be a general theory that would embrace both kinds of entities. But, if there is no such theory, then there cannot be a global category that includes both kinds of entities. A theory, such as Armstrong's, that attempts to reduce all properties to structures seems unable to explain the potentialities and dispositions of things in nature. A theory, such as Shoemaker's or Bird's, which seeks to understand categorical structures in terms of causal powers, or other dispositional properties, seems bound to fail, if only for the reason that it is unable to deal with the problem of location. But, perhaps my own theory fails, because it fails to unify.

Historically, the main reasons for thinking that causal powers and categorical structures belong together ontologically are linguistic and logical. Linguistically, there is the fact that both are commonly predicated of objects, and that such predications may be either true or false, depending on whether the objects in question satisfy these predicates. Logically, n-place relations are universally regarded as n-place predicates, which are interpreted semantically as sets of ordered n-tuples of objects satisfying these predicates. But, in my view, it is a mistake to try to derive ontology from either language or logic. This was the mistake that led to logical atomism in the early years of the last century, and to 'possible worlds' realism in the 1970s. As I have argued elsewhere (2009: 14–22), the truthmakers required for metaphysics are very different from the truth conditions required for logical theory.

From the point of view of a logician, I do not think it matters much what kinds of things predicates are satisfied by. So a distinction between two different kinds of truthmakers for predications would not be disruptive in logical theory. But from the point of view of a metaphysician, there would appear to be too great a difference between categorical structures and dispositional properties for it to be bridged by a unifying theory. Let us therefore speculate that they belong to different ontological categories, one of structures, and the other of dispositions. The category of structures must include all spatiotemporal and numerical relations, and all of the shapes, sizes, orientations, and so on, that are definable as structural relations between things or their parts. This is, more or less, the image of primary reality that was once presented by Locke. But there appear now to be other quiddities beside these. Consider Schrödinger waves. They act as particles, and so might be thought to harbour causal powers, just as Lockean primary structures do. On the other hand, they are transmitted as waves that are unobservable in transit. So, the waves themselves might be thought to provide only the locus of these powers. There is, therefore, a strong analogy between the wave structures of quantum field theory and the Lockean primary structures that were once thought to underlie everything.

It is too early yet to conclude that quantum fields are the quiddities of the world-view that results from quantum mechanics, just as Lockean primary substances were the quiddities of the world-view generated by Newtonian mechanics. But the idea is certainly very tempting.

DEFENDING CATEGORICALISM

D. M. Armstrong

1   Introduction

The three contributors to this symposium have some important metaphysical agreements. We uphold against nominalists the real existence of instantiated properties and relations, which we take to be universals. Perhaps Alexander and Brian would agree with me further that properties strictly so-called are, as Russell thought, simply the monadic case, with relations the dyadic, the triadic ... the n-adic cases of properties in a more extended sense. Again, we all subscribe to the idea expressed with elegant succinctness by David Lewis when he spoke of a *sparse* theory of properties and relations. A monadic predicate, for instance, does not necessarily pick out one of these sparse properties, even in a true proposition. Still further, we agree that it is up to empirical science to tell us just what these universal properties and relations are.

After that the trouble starts. For Alexander the true properties (using the word in its broad sense to cover relations) are dispositional in nature. They have, every one of them, an 'if ... then' structure, a *power* structure. It is what they do in suitable circumstances that constitutes their nature, indeed is their whole nature. For me properties have an intrinsic nature that is not a power. There is an obvious compromise, one that Brian favours: some but not all the true properties are powers.

Some readers may not be aware of the system of voting that prevails in Australia, which is called preference voting. You fill in your ballot giving one candidate a 1, another candidate 2, and so on through all the candidates. A candidate that gets more than half the first preference is elected. But if no candidates achieve this, the candidate who did worst has his number 2 preference distributed to the others. This process goes on, always eliminating the loser, until somebody gets to the magic more than half.

What is the moral for philosophers? It is, I think, that where there are more than two competing theories you should try, if possible, to indicate which theory you would retreat to if your view turned out to be incorrect. After all, what we know about philosophy and its history makes it likely enough that our philosophical preferences are no better than arguable. A wider spread, therefore, might be a good idea. So I announce my preferences in the great metaphysical property election. My ballot paper goes Categoricalism, Middle Way, Dispositionalism.

2 Dispositionalism and the Middle Way

Why do I place Dispositionalism last? I'd agree with dispositionalists that there are certain properties that seem to be rather natural candidates for being powers, for instance mass and charge. It is easy to think of such properties as powers because it is what they do to other things when they interact with them that catches the intellectual eye. But there also seems to me to be candidates for properties such as *distance in space and time, shape, size* and *being in motion* which, even when relativized, I find it hard to see as being nothing but powers. I lack any argument that shows that they *cannot* be presented as powers, but I think that, for instance, the relation of distance constitutes a challenge to pan-dispositionalism that I doubt can be met. So, second preference-wise, I am with Brian rather than Alexander. But even if there are powers, I think that there are also real properties and relations of things that are categorical (a term, by the way, that I got from my supervisor at Oxford, H. H. Price).

Since, as I think, we are stuck with the categorical anyway, it would be a nice simplification if all the true properties are categorical, that is, non-dispositional. The rest of this paper is devoted to trying to prop up this view, my first preference. David Lewis's neo-Humeanism, I take it, also gives a categorical account of properties, but I reject it because it takes too weak a view of laws of nature. The laws are more than just regularities. (For extended criticism see my *What is a Law of Nature?*, 1983.) We need a stronger link between the properties involved in the laws of nature.

3 Defending Categoricalism

I will consider first Robert Black's charge of *quidditism* in his influential article 'Against Quidditism' (Black 2000). The argument is that the categorical properties that I, together with Fred Dretske and Michael Tooley, postulated in order to give an account of laws have no necessary connection with what actually happens to the particulars that instantiate these laws. The argument is pressed by considering a possible world in which the categorical properties are swapped around in some systematic way, but objects continue behaving in just the way they do in our world. This is supposed to refute, or at least to call in question, a categoricalist theory.

It is true, I concede, that if the assemblage of all possible worlds actually exists, with our world only one of them, then this argument would have considerable force. But if one is, as I am, a chauvinist about our world, one who thinks that our actual world is the only world there is, then I think that the force of the argument is greatly reduced. Possibilities that are mere possibilities, that is, possibilities that are not actualities in this world, the only world I think, are not something that a metaphysician should have to take very seriously. There is, let me concede, at least for the sake of argument, no *contradiction* in the supposed swap around of properties that Black (2000) suggests, but that is all. That is a rather weak argument just by itself.

In philosophy we are constantly faced with the necessity for hard choices, one might say the necessity of deciding which is the least unsatisfactory position to take. That is the nature of the beast. And I think at this point that this possible worlds argument of Black's is an argument that it is best to reject. I'm suggesting that we should just concede that it is possible—that it is a non-contradictory supposition—that the postulated contingent connection of universals does not hold, but that some juggled arrangement holds instead, but the way the universals are juggled ensures that the juggling would not be detectable. To this I am saying 'Why not just accept this?' *as a possibility*. There is a good precedent. It is *possible*, we who make the nomic connection of universals a contingent one have always said, that the neo-Humean picture is the true one. It is a possibility we admit, but that proves no more than this: it does not involve a contradiction. But, we maintain, the best *explanation* of the regularities of the world is that it is governed by strong laws of the sort that we uphold. It will be noticed that my reply here to Black's argument is one that has been given a technical name by Daniel Dennett in his *Philosophical Lexicon*. It is an instance of outSmarting, with upper case 'S', named after Australia's own Jack Smart. You meet an argument meant to refute you by accepting its conclusion, but maintaining that the conclusion is no refutation.

Another reply would be to hold that the categorical laws are not contingent but are necessary. But I find such a necessity to be too much *ad hoc*.

4   Laws of nature

In my original treatment of my theory I said that the basic form of a law of nature was a higher-order connection between properties that could be symbolized using a simple and schematic example the law that Fs must be Gs, the connection as N(F, G) with F and G universals, the idea being that this higher-order connection between F and G *entailed* that all Fs are Gs is a regularity in the behaviour of things, a non-supervenient connection justified by its explanatory value. This view was rather justifiably criticized as solving the problem in a somewhat empty manner. There was also a technical difficulty. Like sometimes produces like, most notably when a thing continues to exist. It is plausible to argue, and I think true, that the earlier state of the object *causes* the later stage of the object's existence. (This is Russell's idea that a continuing object is a causal line. See Russell 1948.) Yet how was this to be treated on the connection between universals theory? N(F, F) seemed to be near unintelligible.

I believe that I got round these difficulties in my *World of States of Affairs* (1997). First a word on states of affairs. These are Russell's 'facts' in his *Lectures on Logical Atomism*, and also Wittgenstein's 'facts' in his *Tractatus*. Russell and Wittgenstein suggested that the world was a world of facts, and, apart from vocabulary, I accept this view. The basic states of affairs are such things as a's being F or a's having R to b, with F and R (sparse) universals. For Russell and myself there are further states of affairs, general states of affairs and negative states of affairs, although Russell pointed out later that negative states of affairs can be dispensed with provided you accepted general states of affairs (universally quantified states of affairs). I came to this economy myself without realizing I had been preceded by Russell. Over and above these states of affairs, not supervenient upon them, are causal/nomic connections between the basic states of affairs, connections, new extra states of affairs, that hold in virtue of the universals they involve.

A way to understand these causal/nomic connections is through the notion of a 'state of affairs *type*' introduced in *A World of States of Affairs*. If a's *being F* is a state of affairs then its state of affairs type is _'s *being F*. Many states of affairs may have the same state of affairs type. (If a is an electron and F is its charge the number will be very large, perhaps infinite.) Then we have:

$$\Phi_1\text{'s } being\ F \text{ causes } \Phi_1\text{'s } being\ G$$

This says that something's being F causes it to be G. The subscript shows that it is one and the same particular that is involved.

A slightly more realistic law might be:

$$\Phi_1\text{'s } being\ F\ and\ having\ R\ to\ \Phi_2 \text{ causes } \Phi_2\text{'s } being\ G$$

This says that something's *being F and having relation R* (where R is a spatial relation perhaps) to some further thing, causes that further thing to be G. Then it is asserted that laws of nature are relations holding between these state of affairs *types*. Note that types are all universals.

The difficulty in the case of *like causing like* is then removed. N(F, F) always sounded crazy. But it is easily symbolized using state of affairs types. We have:

$$\Phi_1\text{'s } being\ F \text{ causes (at the next moment, say) } \Phi_1\text{'s } being\ F.$$

Russell's idea that a continuing thing is a causal line could then be represented by means of causal relations between states of affairs types.

It seems to me that this connection is a sort of intermediate necessity for laws between Humeanism and the necessary connection favoured by power theorists. It is contingent but it causally connects states of affairs *types*, which yields a stronger than Humean connection. The universe is more unified, if we make this assumption. (And induction is much easier to justify.)

5 Determinable properties

There is a second difficulty, though. Laws of nature are characteristically mathematical, and connect variables rather than individual states of affairs types. Newton's gravitation law will serve as an example, even though it is now thought to be only a first approximation to the truth. $F = GM_1 M_2 / D^2$ connects an indefinite, perhaps infinite, number of particular cases. In my 1997 book I suggested that we use W. E. Johnson's distinction drawn in his *Logic*, Vol.1 (1964), between *determinable* and *determinate* properties to explain the nature of such laws. *Having mass*, for instance, is a determinable property, while *having mass of 1 kilogram* is a determinate property. Possession of a determinate property entails possession of the corresponding determinable. I had previously held that determinable properties were not universals. It was determinates only that should be accepted as universals. But Newton's law connects determinables, laws were supposed to be connections between universals (more strictly connections between state of affairs types), so it seemed that there was need to accept determinables as universals.

Recent discussions with James Franklin has brought me back, though rather uncertainly, to my original position, that determinables, though properties, are not universals, or at any rate that there is no necessity to admit them as universals. The key point is that actual gravitations are all determined by the determinate values of mass and distance in the actual situation. So it is tempting not to postulate any universals except these determinate values. Then Newton's law becomes a bundle of connections, with the members of the bundle all causal connections holding in virtue of instantiated determinate masses and distances. We could call them *particularized* laws.

An important thing to notice about these particularized laws is that not only do they involve universals, but they are themselves universals. They are complex, not simple, universals, structural universals as I call them, and they involve the universal *causation* (singular causation, not Humean causation) that binds the constituent universals together. As one who rejects uninstantiated universals, I say that these structural universals, the particularized laws, must be instantiated. They are instantiated wherever the causal process is. And this is all that there is ontologically to the particularized law.

Going back to the determinable property (assuming it is not a universal) we can see how the bundling is done. Consider the determinable property of *having mass*. Each determinate stands to each other determinate as includer or included. A pound is a proper part of a kilo, in the sense that for any particular object that is a kilo in mass there are proper parts that are a pound in mass. There is a one-dimensional scale of all and only the mass universals. This appears to be a simple mereological relation that relates each mass-universal to each other that does not call for a determinable universal. The same applies to distance. Note, however, that other laws call for different and perhaps more complex uniting principles.

So my position now is that determinable properties are real properties, but are not universals, Laws like Newton's connect determinable properties. The connection between universals, between state of affairs types, is found in the determinate values only, and it there that suggested laws connecting determinables are verified and falsified. Newton's law was in trouble when calculable cases involving the perihelion of Mercury did not fit the facts.

6   Uninstantiated universals

There leads on to another problem though. May there not be possible determinate universals falling under certain determinables, but with one or more universals that are *not* instantiated? An example might be furnished if the mass of the universe is finite only. This seems to be a possibility. But we might we not think about this situation that the gravitation law ought to apply to greater masses than the mass of the world? So does not my theory need to cover laws that hold for uninstantiated universals?

One way of dealing with this problem would to go a way towards Platonism and postulate universals that are not found instantiated in the natural world. As a naturalist, holding that it is plausible to think that the natural world of space and time is all there is, I would not be happy with this conclusion. I think it is possible to give a counterfactual account of such cases. If these greater masses had existed, then it can be asserted that determinate universals would have been instantiated that in fact are nowhere instantiated. Counterfactuals, of course, must have truthmakers, but it seems that they are readily available. The connection of determinables covering the relevant particularized laws, assuming the laws are true laws, will be the truthmaker. We have the law, the bundle, we have a false premise that certain determinates falling under the law are instantiated, and a false conclusion that certain other determinates that are not instantiated may be drawn. Some may want to postulate uninstantiated universals to be the truthmakers in such cases, but I suggest that provided we have the true bundle of laws then the connection between determinable properties that specify the bundle will be sufficient as truthmakers in such cases. I'd hazard the view that these determinable properties supervene on the particularized laws, and so come at no extra ontological cost.

I have not argued seriously against the positions of Alexander Bird and Brian Ellis. But I have, I hope, said enough to show that a categoricalist about properties can give an account of laws of nature that is at least an arguable view.

7   Kinds

A final note on the metaphysics of kinds. Kinds figure importantly in Brian Ellis's thinking. That particulars can be grouped as kinds of thing is a very important feature of the world. But suppose we are given all the particulars with all their true properties and all their true relations to other particulars. I incline to the view that the kinds supervene, and so are not something ontologically additional to the properties and relations. I'm happy to find agreement with Alexander Bird here.

MONISTIC DISPOSITIONAL ESSENTIALISM

Alexander Bird

According to David Armstrong (1997), all fundamental natural properties are categorical. According to Brian Ellis (2001), this is mistaken, since some properties are essentially dispositional. Brian also thinks that some are categorical and are as David describes them. As regards Brian's first point, I am with him; but I disagree on the second. David is right, all fundamental properties are metaphysically on a par. David is a monist, and so am I; but they are not all categorical, they all have a dispositional character essentially.

1   Laws and the modal nature of properties

The best way of explaining why I think that not all properties can be categorical is to focus on laws. Looking back at the history of my own thinking about laws, I was once a convinced Armstrongian. But then I asked myself, what would the world be like if there were no laws? Indeed could there a world anything like ours but without laws? For David Lewis it would be difficult for there to be a world without laws at all, since some systematisation of the actual non-nomic facts (the arrangement of the Humean mosaic) will be the best system. But perhaps that's not very interesting, since for Lewis laws do not have the metaphysical significance they have for David Armstrong, Brian Ellis, or me. Tea leaves in a cup can leave patterns. Despite what some would-be clairvoyants might think, there is no significance to such patterns. Likewise, from the perspective of Humean supervenience there ought to be no significance to the patterns to be found in the Humean mosaic. Laws are metaphysically much more significant for David Armstrong, and so the question of their possible non-existence is correspondingly rather more revealing. And it would seem that there is no reason why a world could not be exactly like ours in non-nomic respects (i.e. the details of which particulars instantiate which first-order universals), but lack our laws. Such a world would be metaphysically very different, since there would be no second-order relations among universals. And consequently there would be no *explanation* of why particulars possess and change their properties and relations. But such a world would be possible since the possession and changing of properties does not constrain the laws metaphysically—the relationship of metaphysical constraint is the other way around.

On closer inspection I regard such a world as deeply worrying and of dubious possibility. Laws are responsible for the existence and nature of things. Common salt, NaCl, for example is an ionic compound held together by Coulomb force of electrostatic attraction. But no laws means no forces, and so no ionic bonding, and so no salt. I think it is a mistake to think of laws simply as determining what *events* happen or do not happen, or their chances of happening. They also enter into the nature and essence of *things* and *kinds* of thing (Bird 2005c).

But this concern, it might appear, would only apply to complex items and kinds, not to fundamental entities and their properties and kinds. However, it seems to me that related objections do arise. Without laws of electromagnetism, what distinguishes electrons and positrons? In the actual world, the one kind in constrained to behave in different ways from the other kind. But in the lawless world, there are no such constraints. If we go further and remove the law of gravitation and Newton's second law, then we have no distinction between these particles and their related neutrinos; remove further laws and the distinctions between these fermions and other fermions disappear and so on. What it is to be a certain kind of particle is a matter of which laws the particles obey and in which way; remove the laws and one removes the distinctions between the kinds. And that occurs because we lose the distinctions between the *properties* that characterise the kinds. Without laws governing electromagnetism, there is no distinction between positive and negative charge, and between these and having no charge at all.

Arguments of this kind lead me to conclude that there is some metaphysically necessary relationship between laws and natural properties—and of a certain kind: the existence of properties entails the truth of at least some nomic facts concerning them. But they do not tell us how such a relationship arises. Does it arise from the nature of the laws, so that laws are the primary entities, the properties being secondary? Properties would just be 'nomic roles'. I find it difficult to make much of this idea. What then are laws on this view? How do they connect with one another, as laws do? Perhaps further thought will reveal the answer. But for my money a much more palatable approach is the reverse, which takes properties to be the primary entities and laws to be derivative. According to this picture, properties have relationships between them that are essential to the natures of those properties. Laws just fall out as consequences of those relationships. This view is non-atomistic, in the Wittgensteinian sense of 'atomistic'. There are necessary relations between basic entities and so between propositions concerning those entities.

There may be more than one way of articulating the idea that properties have essential relations with one another. My preferred option is to think of properties as having a dispositional nature. This nature, being essential, forges the necessary relations among properties, since, typically, the manifestation and stimulus of a disposition will involve different properties. A property that is essentially dispositional is often called a *power* in the literature, though I prefer the term *potency*, which David uses on occasion, since being less usual in this context is less liable to confusion. ('Power' is often used to denote a something that is, roughly, a disposition with an unspecified stimulus, without any suggestion that such a property is essentially that way.) To rephrase the differences between us, David thinks there are no potencies, and that the fundamental properties are all essentially categorical; I think all fundamental properties are potencies, and none are essentially categorical; Brian thinks that some are potencies and some are essentially categorical.

2 Advantages and disadvantages of dispositional monism

What are the advantages of the view just outlined? And what are the disadvantages and challenges for it?

As is fairly obvious, my view of the fundamental laws of nature makes them all metaphysically necessary. David's view makes them all contingent. Brian's view and mine align in this respect.

I think that the consequence that laws are metaphysically necessary is an advantage, because the thought that the laws are in some way necessary is an appealing and widely held one. The view is of course radically anti-Humean. Fundamental properties are independent existences in the relevant sense: none is a part of any other, nor do they share parts in common (if they did they would not be fundamental). But they do have necessary connections. I think it is possible to show that an attempt such as David's to maintain a Humean conception of properties but nonetheless to have a robust notion of necessity is unstable (see Bird 2005b for the full version of the argument). If that notion of necessity is too robust (i.e. metaphysical necessity), then we have abandoned Humeanism. If it is less robust, then we need an explanation of how it does what it does (i.e. generate regularity). But in examining the resources for that explanation we find that the same question arises. If the explanation has any robustness (i.e. is not mere correlation) then we need some kind of necessity. The latter is either metaphysical necessity or something less. If it is metaphysical necessity, then the position is no longer Humean (we have metaphysically necessary connections). If it is less than metaphysical necessity, we are now a further step along an infinite regress. So Humeanism about properties must be given up. The appropriate notion of necessity for laws is metaphysical.

This does present the difficulty that many also regard the laws of nature as metaphysically contingent. That fact raises interesting questions of methodology in metaphysics. I am willing to regard this intuition as misguided, fallaciously depending on the ability to conceive laws being otherwise. But at the same time, I am willing to employ intuition elsewhere in making my case (e.g. in the initial arguments concerning the essences of kinds and properties). Thus a question is raised, when is intuition to be relied upon and when not? I'm not in a position to answer that question, but here are some thoughts. First, where there is a strong intuition that one rejects one ought to look for an explanation of where it has gone wrong. Intuitions are prima facie plausible, but that is defeasible. Secondly, and moreover, intuitions do not have compelling probative force on their own, but only as parts of a larger theory, which may have other kinds of evidence on its side. Thirdly, one might suggest that intuitions concerning necessity, possibility, and contingency, are, on their own, not especially reliable; more reliable are general intuitions concerning facts about the essence, nature, and identity of things. Then the question is, why should we have especial intuitive understanding of the latter. (The answer is not, of course, that essences are known a priori, though it is interesting that Kripke thinks that while essences are not known a priori, knowledge of the sort of essence a thing or kind has is a priori—as it happens, I don't think that this is always the case.)

One of the principal advantages of dispositional essentialism about properties is that it avoids the problems of quidditism associated with categoricalism about properties. Quidditism is the view that there are no essential differences between properties, at least as far as they interact with one another (e.g. in laws) (c.f. Black 2000). The identity and difference of properties are primitive facts about the relevant properties. Consequently, any nomic role performed by a certain property could be performed by any other property. So there are pairs of worlds that differ simply in the fact that the properties have swapped their nomic roles. An argument, based of one of Chisholm's (1967), suggests that this is implausible. The line of thought outlined above, which takes to be impossible a world in which there are the properties of the actual world but with no laws, also trades on the commitment to quidditism that is present in David's view.

If quidditism is not a plausible view, it is not a plausible view about *any* properties. Rejecting quidditism requires us to reject not only David Armstrong's view, but also Brian Ellis's mixed view. I don't think that spatial and temporal properties are any exception to the rejection of quidditistic, categorical properties. It is, I concede, intuitively attractive to think that spatial and temporal properties and relations are not dispositional, and so at least must be exceptions that favour the mixed view over monistic dispositional essentialism. This is, however, a case where we should be wary of our intuitions.

On the one hand we have the argument against quidditism presented in terms of the question, 'what would differentiate properties in a world with no laws?'. On the other hand, intuition suggests that we can differentiate the spatial relation of 'being 2m apart' from the temporal relation of 'occurring 3s apart' independently of any considerations concerning dispositional features of these relations.

While intuition is valuable, it can be trumped by science. We should reflect on the plausibility of the view that some physicists hold, that a good theory should be background-free (Baez 2000). In classical physics time and space are a background. Space-time is rather like the empty stage before the props and actors are placed upon it. The stage is no part of the action but features in our description of the action ('Enter Benvolio and Mercutio stage left'). Likewise space and time play a part at least in our description of the laws. Yet because they themselves are unchanging, being a mere background, it is difficult to think of their role in the equations as genuinely causal. A generalized action-reaction principle proposes that only what is itself capable of change can be a cause of change. Newton's absolute space-time seems to violate this principle. And so one tempting direction to go in is one we may associate with Leibniz or Kant—space-time is not part of the noumenal world but is part of our framework for experiencing and describing it. However, an alternative is to give up the background conception of space and time, in which case space and time may be genuine patients of change and so genuine agents too. This indeed is the way that the general theory of relativity enjoins us to see space-time. As a consequence it seems possible for spatio-temporal properties to be considered genuinely causal, and hence dispositional.

If all properties are on board as essentially dispositional, we face a problem not faced by Brian, which is the regress problem: since the identity of an potency depends on the identities of its stimulus and manifestation properties, then the claim that all properties are dispositional leads to the worry that there is a vicious regress. Brian stops this by taking some properties out of the equation: the buck stops with the categorical, non-dispositional properties. Instead I espouse and kind of holism or structuralism: the identity of a property is given by the role in plays within the whole structure of properties (Bird 2007b). This can be shown to be non-regressive, if the structure has certain graph-theoretic properties. Appropriate asymmetry in the total set of relations among potencies can ensure the identity of each of them.

3   Natural quantities

How should we understand natural quantities? I have used electric charge as an example, but there are respects in which it is imperfect. (i) the presence of a constant, ϵ_0, in Coulomb's law:

$$\mathbf{F} = \epsilon_0 \frac{q_1 q_2}{r^2}$$

suggests that there could be a similar property governed by a law with a different constant; (ii) electric charge is involved in other laws of nature (e.g. the Biot-Savart law); (iii) possession of a particular value of charge is associated with a multiplicity of dispositions, even if we focus just on electrostatic attraction. None of these observations is inconsistent with the picture outlined above. But they do make it a less clean picture.

Considering (i) would *shmarge*, which obeys an equation like Coulomb's but with a different constant in place of ϵ_0 be the same property as charge? I have some sympathy with the thought that the value of the constant in the law does not determine the identity of the property. In which case there would be some nomic facts (e.g. the value of the permittivity of free space) that would not be fixed by the essences of properties. On the other hand one could have a fine-grained conception of properties and their essences. For various reasons I am inclined to take a platonic view of properties as universals that exist independently of their being instantiated. The fine-grained view would then imply a world full of properties of slightly differing natures, most of which are uninstantiated.

The fact that charge also participates in the Biot–Savart law:

$$d\mathbf{B} = \frac{\mu_0}{4\pi} \frac{(\mathbf{J}dV) \times \hat{\mathbf{r}}}{r^2}$$

(where $\mathbf{J}$ is the current density, i.e. the density of charge flowing per unit time) suggests that charge is a multi-track disposition: one that has a multiplicity of manifestations for a multiplicity of stimuli. Again, one could accept this as a basic fact. But my intuition is that multi-track dispositions ought to be explained rather than posited as fundamental. If one and the same property does more that one thing, then one might wonder whether in fact this property is not basic but is in some way complex, the reflection of simpler dispositions compounded together at a more basic level. Understanding French is a multi-track disposition which we do not think of as fundamental, but in some way or other is compounded out of simpler dispositions.

A similar conclusion may be reached from consideration (iii): possession of a particular charge, say 10mC, confers on its possessor a range of dispositions to exert and to experience a force, depending on the charges and positions of other charged objects. If we think of dispositions as being individuated by stimulus and manifestation pairs, then such properties must be understood as conjunctions of dispositions, not as single dispositions. Or maybe the stimulus-manifestation conception of dispositions needs to be reconsidered: maybe they should be considered more like mathematical functions. (A mathematical total function could be considered as the fusion of many partial functions, one for each value in the domain of the total function. But that would be an unnatural way of looking at things.) However, this does raise questions of the kind we have seen already. For any function there are similar functions in the vicinity. These are also possible properties, and on my platonic view are therefore actual if uninstantiated properties. My intuition is that any relationship between property values as exemplified by Coulomb's law is not basic but stands in need of explanation. That of course is speculative metaphysics. But this speculation is, I think, in tune with the speculations that some scientists are apt to make (c.f. Weinberg 1993). To speculate wildly, the best way the world could turn out to be, for my theory, is a world of on-off properties standing in simple, single-track relations to one another. It would be interesting to model such a world, and to show that a world of apparently multi-track, functionally related properties could supervene upon it.

4   Conclusion

The last few years have seen a great deal of fruitful work on the questions being addressed in this symposium, much of it by David and Brian, while most of rest has been inspired by them. While we have not resolved these big questions, and are perhaps unlikely to do so to everyone's satisfaction, I do think that we have made significant progress in understanding what the key issues are and in developing the arguments that even if not decisive, play a significant role in shaping the landscape. Fruitful philosophical contributions to these debates are still forthcoming. At the same time, our metaphysics must be naturalistic to the extent that its purpose is to contribute to a coherent overall picture of the world, a major component of which is supplied by natural science. While metaphysics should not be enslaved to science, it is nonetheless true that harmony between a metaphysical view and the deliverances of science is a point in favour of that view. And so in adjudicating between our proposals, we should keep an eye on how science, fundamental physics in particular is developing. My own view involves certain implicit bets as to how things may turn out in that science. It will be interesting to see how those bets turn out.

Part II

LEVELS OF INQUIRY

LEVELS OF REALITY AND SCALES OF APPLICATION

Patrick McGivern

Philosophers and scientists often describe theories, laws, and explanations as applying to the world at different 'levels'. This idea of a 'level of application' is often used to demarcate disciplinary or sub-disciplinary boundaries in the sciences. For instance, stoichiometric laws and quantum mechanical laws might be said to describe chemical phenomena at different levels. More generally, the idea of levels is used to distinguish more fundamental laws or theories from less fundamental ones: more fundamental theories are those that apply at more fundamental levels.

In this paper, I will examine this idea of the level of application of a theory, and in particular the idea that the 'fundamentality' of a theory can be understood in terms of the fundamentality of the level at which that theory applies. I'll argue that, in this context, the traditional concept of a 'level' should be replaced by the more general concept of *scale*. Scales and levels are closely related—in fact, the terms sometimes appear to be used interchangeably. But we can distinguish between the traditional account of levels in philosophy and an account of inter-theoretic relations based on scale. I'll argue that the scale-based account is the more appropriate one for understanding science.

There are three advantages that a scale-based account of inter-theoretic relations holds over a traditional levels-based one. The first is that it has greater consonance with actual science: the levels account is perhaps suitable in certain contexts, but the scales-based account is better suited to a wider range of theories in science. The second advantage is that the scales-based account suggests a clearer distinction between a realist account and an anti-realist one; I'll argue for a realist understanding of scales, but the initial point here is simply that on a scales-based account it is clearer what the point of contention is between realism and anti-realism, whereas on the traditional account of levels this is not so clear. Finally, the scales-based account suggests how we might respond to some metaphysically puzzling situations where 'non-fundamental' theories appear to be explanatorily indispensable: they provide explanatory insight that cannot be gained from their more fundamental counterparts. On a levels-based account of theory application, we seem to be forced to accept the idea that, despite the fact that these non-fundamental theories are *false*, they can explain things that no true theory explains; I'll argue that a scales-based account allows us to grant that non-fundamental theories can still truly describe the world, and thus, unsurprisingly, provide us with explanations of parts of it.

This paper is organized as follows. In section 1, I describe the traditional account of inter-theoretic relations in terms of 'levels', focusing specifically on the metaphysical questions of what levels are, what it would mean for them to be real, and what it takes for one level to be fundamental or more fundamental than another. In section 2, I examine similar questions about an account of inter-theoretic relations based on scale. Then, in section 3, I turn to arguments about the relationship between fundamental theories, fundamental levels, and fundamental scales.

1   The traditional view of levels

In this section, I'll describe the traditional account of levels and try to answer three questions about levels:

  i. What is a level? What do we refer to when we talk about 'levels'?

  ii. What would it take for levels to be real?

  iii. What does it mean for a level to be 'fundamental', or for one level to be 'more fundamental' than another?

The clearest statement of the traditional account of levels is that found in Oppenheim and Putnam (1958). The framework of levels is used there as part of a greater argument for the 'unity of science': levels are supposed to allow us to see how all (legitimate) sciences can be seen to be united, both by showing how their domains are related and by showing us how the evidence for 'reductions' between pairs of branches of science provides evidence for the overall reducibility of all science to physics.

On Oppenheim and Putnam's account, levels form a hierarchical structure such that each member of the structure corresponds to the universe of discourse of a distinct 'branch of science'—where branches of science are roughly characterized as collections of theories or laws. The structure of the hierarchy is determined by the part-whole relation, so that members of lower levels compose members of higher levels. However, individual levels are also closed under composition, so that the 'bottom' level, for instance, includes all elementary particles as well as everything composed of elementary particles. Thus moving up the hierarchy doesn't quite correspond to moving from a simpler to more complex domain: instead, each progressively higher level becomes more uniformly complex, though the complex occupants of level six, say, also occupy level one.

I take it that Oppenheim and Putnam's central goal was to articulate the levels framework and to show how it is helpful in thinking about scientific unity. Their specific account of *which* levels there are in the world is perhaps a more tentative suggestion: their suggested six levels range from the level of 'social groups' (level 6) to that of 'molecules' (level 3) to 'elementary particles' (level 1). Whatever the appropriate division into levels—for example, Conger (1925) describes a hierarchy of 25 levels—it's the relationship between individual levels and branches of science that matters most to Oppenheim and Putnam: we could dispute their specific choice of levels without undermining their claims about the basic framework.

With this general framework in mind, let's consider our three questions. First, what are levels? On the traditional account, levels are *sets of entities*: when we refer to a particular level, we are referring to the set of entities that forms the universe of discourse for a particular branch of science. There are further conditions that need to be fulfilled before a collection of sets can be counted as a hierarchy of levels: Oppenheim and Putnam suggest a variety of constraints, including that there be a unique 'lowest' level, that the elements of different levels be related by the part-whole relation, and that the levels 'must be selected in a way which is "natural" and justifiable from the standpoint of present-day empirical science' (Oppenheim and Putnam 1958: 6). But the levels themselves are simply sets of entities.

If levels are sets of entities, what would it mean for levels to be 'real'? I take it that those who dispute the reality of levels, or the adequacy of the concept of levels—such as Heil (2003)—do not specifically mean to dispute the reality of abstract objects, such as sets, in general (of course, they might dispute this as well—but these should be distinct disputes). And yet it does seem legitimate to ask whether levels themselves are real. Levels are often discussed as if they were ontological posits, and while we might reasonably reject the view that levels *are* real, the present point is simply that, given the traditional account, it isn't clear what would be in dispute.

Note that one thing we can't mean by the 'reality' of levels on the traditional view is the question of whether or not science is unified in the way Oppenheim and Putnam suggest: understanding the framework of levels is meant to facilitate the argument for scientific unity, but positing the framework and the argument for unity are distinct.

We could tie the reality of levels to the suitability of the entire framework: in disputing the reality of levels, we might dispute the claim that the universes of discourse of various branches of science can be arranged in a hierarchical manner, converging on a single unifying branch. This is the sort of point that is often made concerning the levels view, as, for instance, when Schaffer (2003a) questions the commitment to a single 'lowest' level. Alternatively, perhaps we should be skeptical of the claim that the domains of all sciences can be put in a single linear ordering: perhaps we should expect branches to form in various directions. But while these points might bear on the question of scientific unity—if, for instance, not only were there no lowest level, but if distinct theories were required for each progressively lower level—they don't seem to be especially important for the question of the reality of the levels themselves. The fact that there could be infinitely many levels, for instance, doesn't give us any obvious reason for thinking that the levels themselves should thus be said to be unreal.

Alternatively, we might try tying the reality of levels to the reality of composite entities in general, in something like the manner of the 'mereological nihilism' described by Rosen and Dorr (2002). In the traditional framework, 'higher' levels are supposed to consist of subsets of lower ones, restricted to increasingly complex entities. If there were no complex entities, then apparently there would not be any higher levels. However, the case isn't entirely clear. Rosen and Dorr's suggestion is that we can replace all reference to complex entities in science with reference to arrangements of simple entities. If this in fact could be done, then presumably we could reconstruct the traditional framework of levels, again as a series of subsets. As in the traditional account, the lowest level would contain all elementary particles; however, in this revised account, it would contain nothing else—since according to mereological nihilism, nothing else exists! We could still distinguish higher levels, though: these would be subsets of the set of all elementary particles—namely the subsets of those particles arranged atom-wise, molecule-wise, cell-wise, and so on.

Leaving the question of what the reality of levels would amount to somewhat undecided, let's turn to the fundamentality of levels, where we can ask two questions: (i) when is one level 'more fundamental' than another? And (ii) what would it take for there to be a 'fundamental' level?

First, note that 'fundamentality' is an intricate concept that admits of several ambiguities. It is typically understood as a relation: we speak of one thing being more fundamental than another. This relation defines an ordering, and fundamentality *simpliciter* can then be understood in terms of that ordering. Again, there are several possibilities. Things arranged according to the 'more fundamental than' relation could be well-ordered, so that any two things in the ordering can be compared in terms of fundamentality; or there could be branching, so that two relative fundamental things could themselves be incomparable. Branches could end in 'most fundamental' things, or the fundamentality ordering could continue on indefinitely.

The concept of fundamentality itself includes both an ontological and an epistemic dimension: the ontologically fundamental is in some sense the 'ground of existence', while the epistemically fundamental is in some sense the 'ground of explanation'. So far, so vague (to borrow from Robin Hendry).

Unfortunately, making clearer sense of fundamentality directly is quite difficult; fortunately, the fundamentality relation itself is typically assumed to track several other, more clearly understood relations: parts are more fundamental than wholes; independent or subvenient properties more fundamental than dependent or supervenient ones; exact theories are more fundamental than approximate ones; exceptionless theories more fundamental than exceptionful ones. If these various orderings—or at least a distinguished subset of them—consistently marched together, then fundamentality itself could perhaps be understood as a composite notion, drawn from both ontological and epistemic components.

On the traditional account of levels, there are several senses in which one level could be considered more fundamental than another. First, 'higher' levels consist of entities that are composed of things belonging to 'lower' levels: lower levels could thus be more fundamental in the sense that parts are more fundamental than wholes. Second, higher levels also form subsets of lower levels, so lower levels could be more fundamental in the sense of being more inclusive that higher levels. And finally, to the extent that successful reductions between branches of science concerned with different levels can be found, lower levels could be more fundamental in the sense that their associated theories make the theories associated with higher levels redundant.

Note that the traditional account involves the claim that lower levels are fundamental in both an 'ontological' sense (parts are more fundamental than wholes) and an 'epistemic' one (theories of lower levels explain everything explained by theories of higher levels). I take it that this combination of ontological and epistemic import is vital for an account of fundamentality in a robust sense: of course, we are free to define fundamentality in any way we like, so we could always simply define it in terms of part-whole relations or whatever. But in that case, 'fundamentality' is simply being used to refer to another relation. It is only when fundamentality involves a combination of ontological and epistemic significance that it becomes a significant concept of its own.

Having described the traditional view of levels, I'll now turn to some examples from physics to illustrate the concept of scale and discuss how ideas about scales and levels relate.

2   The concept of scale

Oppenheim and Putnam's account of levels was presented as part of an argument for the overall unity of science. The primary goal of this argument was to show that all sciences are unified in being reducible in a particular sense to physics. The question of reducibility *within* physics was not considered. However, it's clear that we can raise questions about reducibility and fundamentality within physics that are very similar to those concerning the relationship between physics and other branches of science. So, how well does the traditional account of levels suit cases in physics?

Within physics, distinctions between levels are usually described in terms of *scale*. In many cases, these distinctions seem to map neatly onto the traditional account of levels, as when we distinguish between macro scale theories and micro scale ones: we expect micro scale theories, such as molecular dynamics, to be concerned with entities (molecules) that compose the entities described by macro scale theories, such as hydrodynamics. But there are significant differences between the concept of a level and the concept of a scale. One of these differences is that while levels are understood specifically in terms of the entities described by a theory, scales are associated with the properties of those entities that we try to describe and measure. Scales come in various forms: there are spatial scales, temporal scales, energy scales, and others. Also, unlike in the traditional account of levels, individual theories are not necessarily confined to individual scales: explaining many phenomena requires the use of 'multi-scaling' techniques that combine descriptions on more than one scale at once. For example, figure 1 illustrates the behaviour of a simple physical system—a damped harmonic oscillator—over time. Correctly modelling its behaviour involves both the rapid 'fast-scale' oscillations and the gradual 'slow-scale' decay in amplitude.

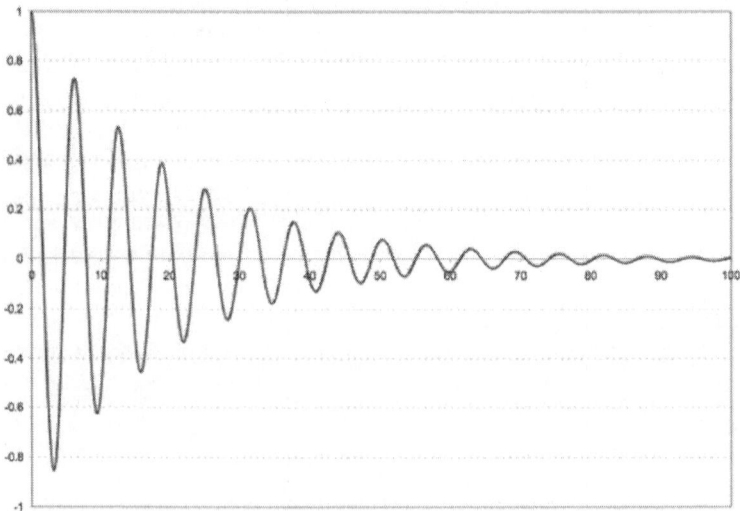

Figure 1

Distinguishing scales in this way is often an important part of describing complex systems such as chemical reactions, biochemical mechanisms, and atmospheric processes. Yet this distinction in scale doesn't correspond to anything in the traditional hierarchy of levels: for instance, the fast-scale behaviour and the slow-scale behaviour are not behaviours of part and whole.

We can try to clarify the concept of scale by considering the analogues to the three questions we asked earlier about levels. First, what are 'scales'?

The term 'scale' is sometimes used almost interchangeably with that of 'level'—phrases such as 'at the micro scale' and 'at the micro level' are seen as being virtually synonymous.[1] But scales are not domains in the way that levels are—though choosing a scale may induce a choice of a domain. Properly understood, scales are better understood in terms of their association with measurement: we measure quantities according to a particular scale. It is a familiar fact that there are different systems of units for making measurements— we can measure distance in feet, inches, meters, etc. In these cases, different scales of measurement can be understood in terms of different mappings from quantities to numerical values: measurement in feet versus inches, for example, gives us a different numerical value for an object's length. But when we consider the scale at which a theory applies, we are interested in more than just a choice of mappings from quantities to numbers: we are also making a choice about the sorts of variations in features considered to be significant.

Philosophers often talk as if 'macro-scale' properties include only what we might call the 'gross' properties of an individual, such as an individual stone's mass or hardness. What seems to makes these properties macro scale properties is that they are instantiated by relatively 'big' *things*, such as tables and diamonds. Conversely, on the standard account, a property counts as a *micro* property just in case it is characteristically a property of a relatively small thing, such as a molecule or atom.

However, this 'entity' oriented view of the distinction between macro and micro scale properties isn't the one used in physics itself. The view from physics allows for properties of 'macro' things to provide much more detailed specifications of a given property than the standard philosophical examples suggest. Macroscopic properties are often best seen as 'structural' properties pertaining to the characteristics of parts of an individual. For example, considering a property such as temperature, we can talk about the *distribution* of temperature throughout a body. We can give a macroscopic description of this property, not simply by attributing the property to a relatively 'large' body, but also by describing that distribution on a macroscopic scale. The scale at which we describe a property in an individual is a measure of the level of detail we are concerned with: a macro-scale representation ignores small-scale variations in a property, while a micro-scale description takes those variations into account. Macro and micro scale descriptions are thus distinguished not so much by the relative 'size' of the things they describe as by the level of detail they take into account. Similarly, when we consider the 'slow-scale' behaviour of the damped harmonic oscillator illustrated earlier, we disregard the rapid

---

[1] This is true even of the broader suggestion from Oppenheim and Putnam that 'levels' should demarcate disciplinary or sub-disciplinary boundaries. For example, in a discussion of 'fundamentality and numerical scales', the physicist Max Dresden defines a scale as a subdivision of physics possessing its own laws and concepts, that is approximately autonomous and 'approximately closed' (Dresden 1998: 476).

oscillations and pay attention only to the gradual long-term decay: choosing a 'slow' scale means choosing to treat only long-term behaviour as significant. And again, when we give a macro scale description of a body of fluid, we describe it in terms that presume that no interesting variations occur below certain spatial intervals. However, our description still in a sense describes those intervals (or at least attempts to—perhaps incorrectly): a solution to the Navier–Stokes equations for fluid flow says *something* about the behaviour of any arbitrarily small region of the fluid.

One consequence of this is that scales are not like 'domains' or 'universes of discourse' in the way that levels are supposed to be: different scales are not like different collections of objects. Instead, different scales are better understood as different ways of describing objects—perhaps the same objects. We can give a fast-scale description of an object—the oscillator—and we can give a slow-scale description; and we can give a multi-scale description that combines both. Similarly, we can give a macro-scale description of an object— a body of fluid, say—and we can give a micro-scale description; and we can again give a multi-scale description that combines both.

If scales are modes of description, then the question of realism about scales might seem easily settled: surely we can *describe* the properties of any object or system we like using an arbitrary unit of measure, and disregard variations in those properties as we please. The choice of scale would then seem like an arbitrary convention. Of course, those descriptions might be more or less precise depending on the choice of scale, but there would seem to be no sense in speaking of 'real' scales.

However, there are at least two ways in which we could argue for the reality of specific scales. First, following Ellis (1968), we could appeal to the simplicity they bring to the laws of nature to justify singling out a distinguished set of scales as 'real'. Second, as I'll describe in the next section, we could appeal to the fact that *applying* our theories to specific systems often involves describing those systems on particular scales. This is often a matter of great convenience—for instance, finding the appropriate scales for non-dimensionalizing a problem often leads to considerable mathematical simplification—but it isn't *only* a matter of convenience: in many cases, the equations we assume 'govern' a particular system simply can't be solved without appeal to the scales appropriate to that system.

Similarly, if scales are modes of description, then the question of fundamentalism about scales might seem equally easily settled as the question about realism: if describing a property of a system on a particular scale involves some measure of our indifference to variations in that property, then scales involving less indifference should be more 'fundamental', in the sense of describing the system as it more truly is. Fundamental scales in this sense should be both ontologically and epistemically fundamental. They should be more ontologically fundamental since *less* fundamental scales—scales involving *more* indifference, as it were—*ignore* features of the system through their indifference; thus scales involving less indifference should approach closer to the system as it truly is. And they should be more epistemically fundamental since presumably any explanatory information that can be got by smoothing over the details—as a description on a less fundamental scale does—can equally well be got by keeping those details intact.

In the next section I'll examine some examples that suggest that this last claim does not follow: theories that apply at less fundamental scales can be explanatorily fundamental, even relative to more 'ontologically' fundamental theories that apply at more fundamental scales.

3   Fundamental scales and fundamental theories

A number of philosophers have argued that we can find examples in physics of situations where laws or theories that are 'non-fundamental' are nevertheless explanatorily indispensable: such laws or theories provide insight into phenomena that cannot be gained, even in principle, through more fundamental laws or theories. For instance, Batterman (2006) describes such a situation, involving the theoretical description of drop-formation, for instance as water drips from a faucet. I won't recount the details of the example (see Batterman's discussion for a clear presentation), but instead will only highlight some of its key features.

The phenomenon of interest in Batterman's discussion is the behaviour of a fluid during drop-formation. The central question is whether the 'more fundamental' theory of molecular dynamics is as explanatorily adequate at describing drop-formation as the 'less fundamental' theory of continuum hydrodynamics. I'll return to this distinction between more and less 'fundamental' shortly: for the moment, it suffices to point out—as Batterman does—that an important difference between the two theoretical frameworks is that continuum hydrodynamics is in a fairly clear sense 'more idealized' than molecular dynamics, since the real bodies of liquid we use it to describe are not in fact continuous fluids.

During drop-formation, a 'bulge' of fluid gradually grows in size and eventually separates from the main body of fluid. In the moments before separation, the forming droplet is connected to the main body of fluid by an elongating and attenuating 'neck'. The features of drop-formation we would like to explain concern the changing shape of the droplet and neck during formation: for instance, how, as the droplet forms, the neck stretches and narrows to a point before the drop actually separates, and whether (and why) drops of different liquids will form in the same shape.

From a hydrodynamic perspective, a dripping body of liquid is characterized as a continuous fluid whose behaviour is governed by the Navier–Stokes equations. This is a macro scale description of the fluid body, where we are indifferent to variations in the fluid conditions—including properties such as rate of flow and surface curvature—below a particular spatial range (say, to variations over distances of less than one nanometer). A solution to the Navier–Stokes equations, given the appropriate boundary conditions, will describe the flow of the fluid—and hence the drop formation—at least as long as the scales appropriate to the system are sufficiently large. Solving these equations is in this case an especially complex problem, since the boundary conditions of the fluid body change as it flows (unlike in simpler flow problems, such as the description of fluid flow in a pipe, or over an airfoil), as, say, the 'neck' linking the forming drop to the main body becomes more and more elongated. Nevertheless, we can give a good description of drop-formation from the hydrodynamic perspective, at least until the width of the fluid neck reduces to the molecular scale.

As the neck width approaches that very narrow limit, solutions to the governing equations approach a 'singularity' where the application of the equations breaks down due to divergences in the speed of elongation of the neck and in the radius of curvature of the growing droplet. At the singularity—at the moment the drop separates—we can no longer apply the Navier–Stokes equations in the standard way. However, we can exploit several features of the situation to treat the problem at the moment of breakup as a scaling problem, where we hope to find (and do find) a description of the drop formation that is in a sense independent of scale: at that moment, the problem becomes a 'self-similar' one, and this independence of scale leads us to expect similar patterns of formation when we consider drops of all sizes. Essentially, what happens in this situation, from the macro scale perspective, is that the initial development of a drop depends on features local to the system—the radius of the faucet, for example. This develops relatively smoothly until the moment of separation, when the system is in a sense suddenly transformed into one where those local features are irrelevant; after the drop forms and separates, the Navier–Stokes equations can then be applied to the system again (though now we will be dealing with two separate applications to two distinct bodies).

That's the story from the macro scale perspective—the perspective where we disregard the details of the structure of the liquid below a certain unit of measure. The central point is that despite the breakdown in the application of the macro theory, we find a satisfying, non-dimensionalized, account of the shape of drops at the moment of separation.

From the micro scale perspective, the story is quite different. At this scale we are still indifferent to certain details—we ignore details of molecular structure and treat molecules as individual units—but the measure of our indifference is much finer. The micro scale characterization of the fluid is given in the framework of molecular dynamics—and we can look to simulations of the motions of molecules to find a micro scale account of drop-formation. As Batterman explains (and illustrates) however, the micro scale lacks a unified account of the formation of droplets: we can explain the motions of molecules, but we find no general explanation of the shape of drops at breakup, corresponding to the scaling explanation found at the macro scale. Similarly, we find no micro-scale correlate to the breakdown at the macro scale that led to this scaling explanation: the process that at the macro scale appeared to involve a deep theoretical discontinuity as we approached the singularity, proceeds at the micro level with no indication of a significant theoretical break.

What's more, even at the micro scale—or, more precisely, at the *nanoscale* of the molecular jets Batterman considers—it is the *macro* scale theory that is needed for insight into the process of drop-formation. Nanojets form droplets, too, and it is only by applying the macro theory—treating those nanojets as if they were continuous bodies of fluids, rather than treating them as collections of discrete molecules—that we can find a satisfying theoretical account of this drop-formation.

What are we to conclude from this example? I'll quote Batterman's conclusion at length, because I think it is plausible and yet puzzling. Batterman (2006) concludes that we must distinguish between two senses of fundamentality:

> ...On the one hand, there is a clear sense in which the molecular theory is more fundamental than the fluid dynamical theory. It is more fundamental because it more accurately describes the genuine nature of the system. Fluids *are* composed of a finite number of molecules, and when we look at fluids around the scale of intermolecular distances, that theory most correctly characterizes the system of interest. This is a sense in which the molecular theory is *ontologically* fundamental. It gets the metaphysical nature of the system right.
>
> On the other hand ...such ontologically fundamental theories are often explanatorily inadequate. Certain explanatory questions—particularly, questions about the emergence and reproducibility of behavior—cannot be answered by the ontologically fundamental theory. I think that this shows, that for many situations, there is an epistemological notion of 'fundamental theory' that fails to coincide with the ontological notion.

So, in the present case, the epistemically more fundamental theory is found in continuum hydrodynamics: that theoretical framework isn't simply more manageable (perhaps more 'pragmatically' fundamental) than the framework of molecular dynamics—it is actually explanatorily indispensable, providing explanatory insight that is unavailable from the molecular theory. Nevertheless, it is the molecular theory that 'gets the metaphysical nature of the system right'.

But how can this be? How could a false theory—a theory that gets the metaphysical nature of the system *wrong*—explain features that are inexplicable by a theory that gets that system's nature right? It's easy to see how a false theory might explain some—even many—of the *same* things that the corresponding true theory explains; and it's certainly easy to see how a false theory might 'approximately' explain features most accurately explained by a true theory. But that's not the suggested situation here. Here we have a supposedly false theory that explains things that the corresponding true theory cannot.

It's important to recognize the real source of surprise here: it isn't simply that an ontologically fundamental theory—a theory that gets the metaphysical nature of the system right—might not be explanatorily adequate for some purpose. We might naturally expect ontological fundamentality and explanatory insight to go hand in hand, but I don't think there's any deep reason to think that they *must*: certain phenomena might just be inexplicable. What is surprising in cases like Batterman's is that not only does the ontologically fundamental theory not provide an explanation for a given phenomenon, but some *non*-ontologically fundamental theory—a theory that gets the metaphysical nature of the system *wrong*—does.

I want to suggest that the lesson we should draw from this example, and others like it, is not just that fundamentality is ambiguous between 'ontological' and 'epistemic' senses: 'ontological' fundamentality is itself ambiguous. Recall that earlier I vaguely characterized ontological fundamentality in terms of the 'ground of existence'. But it is important to recognize that this 'grounding' concerns not only what entities there are—atoms vs. continuous fluids, say—but the ways those things behave. And behaviour is something that in general is associated with specific scales, as, for instance, in the distinction between the slow and fast scale behaviours of the oscillator described earlier. So while the molecular theory might be ontologically fundamental in the sense of making the right assumptions about the entities involved, it isn't ontologically fundamental insofar as their behaviours are concerned.

To illustrate this idea using a different example, consider a textbook application of 'multi-scale analysis' in the description of steady state heat conduction in a one-dimensional rod.[2] On the micro-theoretic account, the rod is assumed to be composed of atoms separated by empty space. Conductivity on this view is discontinuous, and varies rapidly with position. On the macro view, on the other hand, the rod is assumed to be homogeneous, and conductivity is continuous, without the rapid variations characteristic of the micro description. The general equation for steady-state thermal conductivity is:

(1) $(d/dx) \, [k(x)dT(x)/dx] = 0$

We can read this equation as expressing a relationship between structural properties characterizing the distribution of temperature (specified by T) and conductivity (specified by k) in the rod. Note that in equation (1) neither k nor T is inherently micro-structural or macro-structural. Instead, equation (1) gives us a schema for representing relationships between various properties: the properties themselves are represented by particular solutions to (1), and it is the nature of a particular solution that determines whether it represents a macro or a micro structural property. A particular solution for T, for example,

---

[2] See, for instance, Holmes (1995: 224ff), Rueger (2006), and Rueger and McGivern (2010).

will designate a micro-structural property if it takes into account the micro-level structure of the body, and a macro-structural property if it 'smoothes over' these fine details. Formally, we can distinguish between these two levels of description in terms of the way we characterize conductivity, so that a function T(x) will represent 'macro' temperature if it is a solution to (1a):

(1a) $(d/dx)[k_{Macro}(x)dT(x)/dx] = 0$,

where k_{Macro} represents conductivity in a continuous 'macroscopic' way; and a function T(x) will represent 'micro' temperature if it is a solution to:

(1b) $(d/dx)[k_{micro}(x)dT(x)/dx] = 0$,

where k_{micro} represents conductivity in a discontinuous 'microscopic' way. It's important to keep in mind that while k_{Macro} and k_{micro} both represent conductivity, they are distinct functions. Each is characterized by a distinct spatial scale, which we'll call L (the macro scale) and l (the micro scale).

Despite the superficial similarity between these two equations, the equation describing the micro-theoretic treatment (1b) turns out to be much more difficult to solve, due to the manner in which thermal conductivity fluctuates on the micro-scale. To find a solution to (1b), we need to apply a perturbation method, where we attempt to find an expansion of the solution that adequately approximates the full solution after only a few terms. Our goal, that is, is to find an expansion of T(x):

(2) $T(x) = T_0(x) + \epsilon T_1(x) + \epsilon^2 T_2(x) + \dots$

around some appropriate leading term T_0, where the coefficient ϵ is a small parameter representing the ratio of the length of the rod to the period of the atomic lattice. Our hope is that considering such solutions in the limit as ϵ tends to 0 will give us an accurate depiction of the system on the micro scale.

Unfortunately, it turns out that this approach doesn't work: the perturbation solutions (2) diverge in the limit, and we aren't left with any accurate account of the system's micro scale temperature profile. This is where 'multi-scaling' enters the picture. The intuitive idea—often described in compellingly realistic terms by physicists—is that we have failed to take into account the fact that there is more than one scale of relevance to the system. Mathematically, the technique of multi-scaling involves introducing new scales—a new spatial scale, in this case—and treating the system as if its properties varied independently on those scales. So, in the case of the rod, we assume that the variable x represents the microscopic scale and introduce a new variable $\zeta = \epsilon x$ representing the macroscopic scale. We then perform the perturbation expansion in terms of the two variables, treating them independently. The perturbation expansion of T is then:

(3) $T(x, \zeta) = T_0(x, \zeta) + \epsilon T_1(x, \zeta) + \epsilon^2 T_2(x, \zeta) + \dots$

Now we use this expansion as our solution to the original equation (1b).[3] The mathematical benefit of this is that we are able to 'control' the behaviour of the expansion terms T_0, T_1, etc. and prevent the solution from wildly diverging, as it did in the original attempt.

The interesting outcome of all of this is the manner in which those terms get controlled: to get a well-behaved solution, we are first forced to choose coeffecients so that the leading order of the expansion satisfies the constraint:

(4) $\partial/\partial x[k(x)\partial T_0(x,\zeta)/\partial x] = 0$

This gives us the 'core', first-order approximation of T. To improve this approximation, we examine higher orders in the expansion. To keep the second order expansion asymptotic (that is, to prevent it from diverging unacceptably from the value it is supposed to be approximating), we need to impose a further constraint on T_0:

(5) $\partial/\partial\zeta[k_{eff}\partial T_0(\zeta)/\partial\zeta] = 0$

where k_{eff} is the 'effective' conductivity:

(6) $k_{eff} = [\frac{1}{l}\int\frac{dx}{k_{micro}(x)}]^{-1}$

The effect of this constraint is to make k_{eff} a *macro* variable: it is indifferent to the small scale variations in conductivity (k_{eff} is the harmonic mean of the microscopic conductivity). So k_{eff} is k_{Macro} mentioned in (1a), and since T_0 satisfies (5), it can be regarded as the *macroscopic* description of temperature!

What this means is that to find a solution for the micro-level description of the relationship between temperature and conductivity in the rod, we've had to use a perturbation expansion, and to keep that expansion asymptotic (and thus to keep it a good approximation), we've had to posit two independent scales on which the properties we're interested in—temperature and conductivity—can vary, *and* we've had to constrain the first component of that expansion so that it represents the *macro* description of the relationship between temperature and conductivity in the rod.

It's important to understand that we didn't *assume* that the micro scale description could be given in terms of the macro—that relationship fell out of the application of the technique of multi-scaling. But the guiding principle for *that* was simply a desire to keep the expansion of the micro scale description well-behaved.

Now consider the question of the fundamentality of the macro and micro theoretic descriptions. Surely the macro theoretic description is wrong (and less fundamental) in the sense of its commitment to the entities composing the rod—the rod is not a continuous substance, and conductivity is in fact

---

[3]Or, rather, we use this expansion as our solution to the multi-scale equivalent of (1), where ordinary differentiation is replaced by partial differentiation to accommodate our multiple independent scales.

not continuous throughout. But the macro-theoretic *behaviour* is an essential component of the micro-theoretic description: it forms the leading term in the expansion, and we were led to build the expansion around that term through a process that only insisted that we kept the expansion well-behaved. And while the choice to include the rod's behaviour on this scale doesn't exactly simplify the *statement* of the relevant laws of nature (those given—we'll assume—in (1a) and (1b)), it does greatly simplify their application—so much so, that it seems we should have every reason of any indispensability argument to suppose that those scales are real.

So while the 'levels' view of inter-theoretic relations encourages the idea that the question of the fundamentality of levels only concerns the entities involved, a 'scales-based' view suggests that we can distinguish fundamental behaviours as well. In both the simple case of the rod and in Batterman's more complex case of drop-formation, the fundamental behaviour of the system arises from the macro theory, rather than the micro theory. When we apply the continuum theory—whether to 'macro' fluids such as dripping faucets or to 'micro' fluids flowing from nanojets—we make use of a theory that is in one sense ontologically non-fundamental: again, it makes assumptions about the entities involved we think are false. But in another sense, it is the more ontologically fundamental theory: it describes the unity of the behaviours of those entities in a way the molecular theory does not. In fact, Batterman's case gives us an example of something we might take to be even more deeply fundamental than either sense of ontological fundamentality I have mentioned so far. The scaling solution that provides the account of drop-formation isn't just an example where a description of behaviour at a scale seems so indispensable as to warrant regarding that scale as 'real'—it's an example where the relevant behaviour is in a sense independent of scale: explanatory unity is provided by the fact that the same scaling solution applies to a great range of scales (including faucets and nanojets). In that case, it seems more appropriate to assert the ontological fundamentality of the behaviour, independently of the fundamentality of a particular scale: scaling phenomena are in this sense the most fundamental phenomena there are.

Part III

REALISM IN MATHEMATICS

ANTI-NOMINALISTIC SCIENTIFIC REALISM: A DEFENCE

Stathis Psillos

1   Introduction

Philosophy of science proper has been a battleground in which a key battle in the philosophy of mathematics is fought. On the one hand, indispensability arguments capitalise on the strengths of scientific realism, and in particular of the no-miracles argument (NMA), in order to suggest that a) the reality of mathematical entities (in their full abstractness) follows from the truth of (literally understood) scientific theories; and b) there are good reasons to take certain theories to be true.[1]

On the other hand, arguments from the causal inertness of abstract entities capitalise on the strengths of scientific realism, and in particular of NMA, in order to suggest that a) if mathematical entities are admitted, the force of NMA as an argument for the truth of scientific theories is undercut; and b) the best bet for scientific realism is to become Nominalistic Scientific Realism (NSR) and to retreat to the nominalistic adequacy of theories.

In what follows, I will try to show that anti-nominalistic scientific realism is still defensible and that the best arguments for NSR fail on many counts. In Section 2, I will argue that there are good reasons not to read NMA as being at odds with the reality of abstract entities. In Section 3, I will discuss what is required for NSR to get off the ground. In Section 4, I will question the idea of the nominalistic content of theories as well as the idea of causal activity as a necessary condition for commitment to the reality of an entity. In Section 5,

---

[1] The best defence of this argument is by Colyvan (2001).

I will challenge the notion of nominalistic adequacy of theories. In Section 6, I will try to motivate the thought that there are mixed physico-mathematical truthmakers, some of which are bottom-level. Finally, in Section 7, I will offer a diagnosis as to what the root problem is in the debate between Platonistic Scientific Realism and NSR and a conjecture as to how it might be (re)solved.

2   From scientific realism to Platonism

A central argument for the reality of mathematical entities comes straight from the philosophy of science. The indispensability argument (*IND-A*), put in a nutshell, suggests that the existence of abstract entities follows from the truth of scientific theories. There has been a lot of discussion about the exact formulation of *IND-A*, but I take it that the simplest and most explicit way to think of it is this:

> *IND-A*
>
> 1. If they are true, scientific theories (taken literally), imply the existence of abstract entities (numbers, sets etc.).[2]
> 2. Scientific theories are true (weaker: there are good reasons to believe that scientific theories are true).
> *Therefore*
> There are abstract entities (weaker: there are good reasons to believe that there are abstract entities).

Seen that way, the claim of indispensability does not come into the argument directly, but is part of its defence. 'Indispensability' supplements the Quinean criterion of ontic commitment. The latter, as is well known, has to do with the values of the variables in the canonical notation of quantification of a theory. Roughly put, a theory is ontically committed to whatever its variables of quantification range over. But then to go from ontic commitment to what there is, *truth* is required. What a theory is committed to is one thing, what there is, another.[3] The two are linked by the truth of the theory. But before any ontic commitments are being read off the theory, the theory should be taken at face-value. More importantly, if the entities which a face-value reading implies are dispensable, that is, if the theory can be re-written (paraphrased, as Quine would put it) without implying commitment to such entities and without losing in explanatory power etc., the question of commitment becomes moot. So: part of the defence of *IND-A* is that a) quantifying over (abstract) mathematical objects is indispensable in science; and b) theories which quantify over math-

---

[2] This, of course, assumes standard semantics for mathematics, according to which mathematical theories—literally understood—refer to abstract entities.

[3] Here is what Quine (1980: 15) says: 'We look to bound variables in connection with ontology not in order to know what there is, but in order to know what a given remark or doctrine, ours or someone else's, says there is'.

ematical entities (as well as physical entities—like electrons etc.) are highly confirmed; hence (likely to be) true.

What needs to be stressed (for the purposes of the issue at hand) is the general idea behind *IND-A*, viz., that scientific realism implies Platonism—or at least that it contradicts nominalism.[4] Concomitant to this general idea is an epistemological *bonus*. If the truth of scientific theories can be known, there can be knowledge of abstract entities too. Perhaps, there can be knowledge of mathematical truths, too—though not necessarily of truths of *pure* mathematics.

The epistemic optimism characteristic of scientific realism is based on NMA. The argument, roughly put, is that empirical success (suitably regimented so as to include novel predictions and the like) offers good reasons to believe in the truth of theories, since it is best explained by the claim that theories are true. Thus conceived, NMA is blind to a distinction between abstract entities and concrete ones insofar as commitment to both types is implied by the truth of (literally understood) scientific theories. In all its generality, NMA does not demand a *causal* explanation of the empirical success of theories. In particular, it does not demand that the entities that are required for the explanation of certain empirical phenomena should be concrete physical entities. This is as it should be, for at least two reasons that are rarely noted. *First*, any serious (let alone the best) explanation of at least some empirical/predictive successes will have to cite/employ natural laws, and these are in no sense concrete entities. Nor are laws causal entities. It might well be that c causes e via a law (i.e., c is nomologically sufficient for e), but laws themselves do not cause anything, though they are explanatory of what happens in the world. (Newton's law of gravity does not cause things to fall to the ground, although it governs—arguably—their relevant behaviour; nor do Newton's laws cause Kepler's laws and the like.)[5] *Second*, any serious (let alone the best) explanation of at least some empirical/predictive successes will cite/employ *models*. These are not concrete causal entities, and yet they play an important explanatory role. Hence, it is a good thing that NMA is not tied exclusively to causal explanations of the successes of theories.

In its usual formulations, the indispensability argument is cast in terms of confirmation of scientific theories and a standard worry (dressed up as an objection) is that confirmation cannot reach the abstract objects posited by theories and the claims made about them. This objection is flatly question-begging. It presupposes that the relation of confirmation mirrors (or ought to mirror) causal relations; that only whatever is concrete and causal is such that assertions about it can enter into relations of confirmation with obser-

---

[4]The stronger claim, strictly speaking, follows only if there is further argument to the effect that abstract entities should not be understood as mental constructions. But I take this for granted in this paper. For some interesting discussion of reductionist versions of nominalism, see Heck (2000).

[5]Even so-called causal laws (e.g., smoking causes lung-cancer) are not causal entities (they have no causal powers themselves), though they capture, arguably, causal relations among properties.

vational reports. If this presupposition were to be granted, universal law-like statements would not be confirmable, since if true, they are made true by entities which are not concrete and causal. Nor would any hypothesis involving theoretical models ever be confirmable.

If *IND-A* is sound, there are mixed facts, viz., structured entities, whose structure is made up by concrete *and* abstract elements. More specifically, some mixed statements implied by scientific theories (of the form 'P is (or has the property) M', where 'P' is meant to refer to a physical object and 'M' to a mathematical one) are made true by mixed facts: mathematical entities (and properties) are somehow related to physical entities to render mixed statements true. *IND-A* need not imply that there is causal contact (or causal glue) between physical entities and mathematical entities that make up the mixed fact.[6] What *IND-A* does imply is that mathematical entities are required for the truth of scientific theory. This should not be confused with the stronger claim that they are *causally* required for their truth. (We shall return to this issue in Section 6.)

3   From scientific realism to Fictionalism

Resistance to Platonism has also found its springboard in the philosophy of science. Hartry Field's (Field 1980) argument against Platonism has been based on the claim that mathematics is dispensable in science, viz., that scientific theories need not employ vocabulary that purports to refer to mathematical objects; hence, the truth of (suitably reformulated) scientific theories does not imply the reality of abstract entities. This is not an argument *for* fictionalism, viz., the view that there are no mathematical objects, but it paves the way for it. What is normally added to get fictionalism is the general idea that there is something *deeply* suspect with alleged abstract entities. That is, the very idea of abstractness eradicates any ontological pedigree. Concomitant to this general idea is an epistemological liability, viz., that precisely because of their abstractness, alleged abstract entities would be unknowable.

A standard argument against Field's anti-platonist move is that scientific theories (especially high-level ones) resist nominalization: they resist a nominalism-friendly reformulation that implies commitment only to concrete entities. This is an empirical matter in that one has to sit down and sketch, at least, how a nominalism-friendly reformulation of *each and every* scientific theory can be achieved. There is no reason to pause over it now, since there is a general strategy for bringing together scientific realism and mathematical fic-

---

[6]This is not particularly surprising. If we take facts seriously as structured entities, (where a property is attributed to an object or a relation is said to hold between a number of objects) the relation that 'holds together' the elements that make up a fact is not causal; sometimes it is said to be formal (e.g., by E. J. Lowe) precisely in order to make this point clear. In a (different but related) sense, *any* fact is a mixed entity (because it comprises particulars and universals); and what binds all these constituents into a single entity is *not* causation.

tionalism, akin to the one Bas van Fraassen (1980) has pursued in the scientific realism debate. This is to introduce the concept of nominalistic adequacy and to argue that even if scientific theories cannot be nominalised, even if mathematics is theoretically indispensable, there is a way to avoid commitment to mathematical realism. It is enough, it is argued, for the applicability of scientific theories and for the explanation of their empirical successes that they are nominalistically adequate, where a theory T is nominalistically adequate if [and only if] 'it is correct in its nominalistically-stated consequences (i.e., if it is correct in those of its consequences that do not quantify over mathematical entities' (Leng 2005b: 77).

Though he does not endorse mathematical fictionalism, Gideon Rosen (2001: 75) has characterised n-adequacy modally thus: A (mathematised) theory S 'is nominalistically adequate iff the concrete core of the actual world is an exact intrinsic duplicate of the concrete core of some world at which S is true—that is, just in case things are in all concrete respects *as if* S were true'. A concrete core of a possible world W is 'the largest wholly concrete part of W: the aggregate of all concrete things that exist in W'.

We shall discuss this concept in some detail in Section 5, but for the time being let us focus on the general idea behind it, viz., that a theory can be *false* (if literally understood), and yet get everything right vis-a-vis whatever is concrete. We tend to forget that the consensus over the claim that scientific theories should be taken at face-value (they should be understood literally) is a hard-won one, fought against reductive strands in empiricism and syntactic instrumentalism. Literally understood, scientific theories purport to refer not just to unobervable entities (over and above observable ones), but also to mathematical entities too. In fact, as noted already, theories typically comprise a host of *mixed* statements, not just connecting the theoretical and the observational vocabulary but also connecting both vocabularies with a mathematical vocabulary. Mixed statements, under the assumption of a literal understanding of them, require mixed truthmakers or mixed facts. If, as it happens, some part of the required truthmaker is missing (as will be the case if there are no mathematical objects), two options are available to a would-be scientific realist. One is to go for a non-literal understanding of the mixed statements, thereby claiming that the *appropriate* truthmaker in not really mixed and hence that no part of it is *really* missing. The other option is to insist on a literal understanding of theories and to concede that they are false. Leaving the first option to the one side, the second option would be *prima facie* disastrous for realism, at least as an epistemic thesis. How, for instance, can this systematic and symptomatic falsity of theories explain their empirical successes?

There is a nominalism-friendly way out of this problem, but it requires that there is a way to:

a) carve up entities into two disjoint sets—the concrete and the abstract;

b) disentangle whatever the theory asserts about concrete causal

entities (its nominalistic content) from whatever it asserts about
abstract ones; and
c) show that whatever credit accrues to the theory from its applications to the world comes exclusively from its nominalistic content.

Assuming that (a) can be dealt with, what Mark Balaguer (1998: 130) has
called 'nominalistic scientific realism' (NSR) aims mainly to deal with (b)
above. He enunciates the following two theses:

(NC) Empirical science has a purely nominalistic content that captures its 'complete picture' of the physical world.
(COH) It is coherent and sensible to maintain that the nominalistic content of empirical science is true and the platonistic content
of empirical science is fictional.

(NC) asserts precisely what needs to be shown. But the argument for it is
just the fact that mathematical entities, if they exist at all, are causally inert (cf.
1998, 132). Hence, there would be no *causal* difference in the world, if they
did not exist. Hence, there is a way the world is—causally—which is independent of any mathematical objects, which are causally inert anyway and hence
cannot contribute to the way the world is causally. The nominalistic content
(n-content) of the theory, then, is what the theory says about whatever is part
of the causal structure of the world. (COH) follows rightaway and makes possible the claim that though, literally understood, scientific theories are false,
it is enough for scientific realism to get right the nominalistic content of theories, since we do not lose 'any important part of our picture of the physical
world' (1998: 134). As Balaguer puts it, 'The nominalistic content of a theory T
is just that the physical world holds up its end of the 'T bargain', that is, does its
part in making T true' (1998: 135). In the end, mathematical fictionalists do not
have to replace platonistic scientific theories with nominalistic ones. They just
need to argue that when these theories are accepted, we are committed only to
the truth of their nominalistic consequences and not to the truth of their platonistic consequences. In other words, scientific realism implies commitment
to the nominalistic adequacy of theories.

Note that the move from 'no causal difference' to 'no difference', which is
required for the assumption that the causal image of the world is the *complete*
image of the world, is fallacious. It would imply that laws of nature make no
difference since they make no causal difference. But laws do make a difference, even if it is not causal. They are unifiers; or they govern/explain their
instances; or (more importantly), being patterns under which sequences of
events are subsumed, they *constitute* what we call the causal structure of the
world, viz., they make causal happenings possible—where this relation of constitution is not, of course, causal. Laws, to repeat, do not cause anything to
happen; laws just *are* and things happen in conformity to them.

This is not yet an objection against NSR, though it will be extended to one
in the next section. For the time being, let us note that even if Balaguer's strat-

egy were impeccable, there would still be need for an argument for part (c) of the tripartite strategy for NSR noted above. A form of this has come from Leng (2005). Her view is that the best bet for scientific realism is to go for NSR—the (alleged) platonistic content of theories is an extra burden that scientific realists cannot discharge. Her argument is this. If true, theories (literally understood) imply the existence of both unobservable *and* mathematical entities. But the mathematical entities are causally inert; hence, they won't be involved in any causal explanation of certain empirical successes of theories (e.g., a novel prediction). Hence, the no-miracle argument that scientific realists employ to ground their epistemic optimism no longer offers reasons to believe in the full truth (nominalistic + platonistic) of scientific theories. And yet, it could give us reason to believe that successful theories are nominalistically adequate. *Ergo*, scientific realism is in much better shape if truth is replaced by nominalistic adequacy.

4   On the nominalistic content of theories

Drawing a sharp distinction between the concrete and the abstract is notoriously difficult, even if there are paradigmatic cases of entities that are concrete and entities that are abstract (see, for instance Dummett 1991: 239). Most typical criteria for this distinction (lack of spatio-temporal location and causal inertness) admit of interesting (though occasionally contentious) counterexamples. In any case, there is little doubt that mathematical objects count as abstract, if only because they are the paradigmatic cases of causal inertness.[7]

Be that as it may, there is a whole category of abstract objects whose existence is contingent (i.e., they do not exist necessarily) and also contingent upon the existence and behaviour of concrete objects. Examples of such objects are the Equator, the centre of mass of the solar system, or even thoughts (if dualism were right).[8] More importantly, abstract objects are the truthmakers of the descriptions of (most) theoretical models employed by theories. The Linear Harmonic Oscillator (LHO), for example, or the two-body Newtonian system, or a frictionless inclined plane are pertinent examples. It's tempting to conflate models with their descriptions. But if care is taken to draw this distinction, models are abstract objects that satisfy certain descriptions. They are not *pure* abstract entities since physical properties are ascribed to them, but they are abstract nonetheless—and certainly not causally efficacious. We

---

[7] Note that *IND-A* is an a posteriori argument for the existence of abstract mathematical objects.

[8] Since this paper was drafted, an important book has appeared by Wetzel (2009), in which she thoroughly defends the idea that there are plenty of non-mathematical abstract objects (notably: semantic types, but also structural types, like flags, sonatas and molecules), which are explanatorily indispensable and not amenable to nominalistic paraphrase.

can borrow Dummett's expression and call models 'physical abstract entities'.[9] Similarly, objects such as the Equator or the first Meridian are geometrical abstract objects.[10]

We may draw a distinction between Non-Mathematical Abstract Objects (NMAOs) and Mathematical Abstract (MAOs). Literally understood, theories imply commitment to a host of NMAOs; that is, to a host of causally inert entities. It is absurd to say that all these NMAOs are not explanatorily relevant to the successes of theories; nor that they contribute nothing to the explanation of concrete physical objects and their behaviour. A LHO, for instance, does explain why the period of a concrete pendulum is proportional to the square root of its length; it supports certain counterfactuals (e.g., about changes of the length of the pendulum); it unifies under a type a variety of resembling concrete objects. It follows that causal inefficacy is no reason to deny that some entity is part of reality. Causal inertia does not imply explanatory inertia.

A *prima facie* plausible riposte available to NSR is that all these entities are dispensable. But this reply would be too quick. Let us distinguish between two types of NSR: Lenient NSR and Austere NSR. The lenient version is tough on mathematical objects, but does allow non-mathematical abstract entities. The austere version puts a ban on anything abstract. The austere version should aim to dispense with all putative abstract objects by reformulating scientific theories so that they do away with them. I do not know whether this is feasible, but suppose it is. The result of this Herculean operation will be, in all probability, a massively complicated theory which would be unable to make any general claims about concrete objects. Generality requires abstractness: otherwise the general cannot cover the particular. There is not a scientific theory of concrete springs, and another of concrete pendula and another of ...: there is a theory of the linear harmonic oscillator, which covers many concrete structures that are inexact tokens of the linear harmonic oscillator.

The lenient version of NSR has, at least, the resources for the development of simple, explanatory and unified scientific theories via representational devices that employ NMAOs. The latter, among other things, provide the resources for the formulation of comprehensive laws. If NMAOs are (allowed to be) part of the content of scientific theories, the very idea of a sharply delineated nominalistic content of a theory that bans abstract entities altogether becomes otiose. For NMAOs (including laws) play a key role in specifying what the theory asserts about concrete objects and their behaviour. They also play a key role in explaining the behaviour of concrete objects. What is more, part of the identity of some NMAOs (more particularly, of models) are mathematical entities, like phase spaces, vector spaces and groups (cf. French 1999). And since NMAOs are, after all, abstract entities, there is no principled prob-

---

[9] It bears stressing that the qualification 'physical' is meant to imply that some abstract objects are described by physical (as opposed to mathematical) predicates. Types (e.g., TIGER or GRIZZLY BEAR, or Beethoven's FIFTH SYMPHONY) are abstract objects of this sort (cf. also Wetzel 2009).

[10] For more on this, see my (2011).

lem in having *mathematical* abstract objects as part of their constitution. So if NMAOS are explanatory, so are those mathematical entities that are part of their constitution.

Friends of lenient NSR might claim that the employment of descriptions that, taken at face value, refer to alleged abstract objects are purely descriptive and representational devices which, though expressionally and theoretically indispensable, are *not* metaphysically indispensable. Here again, however, the only general (and initially plausible) argument for the alleged metaphysical dispensability of abstract objects comes from their causal inertness, and this is not enough to deny existence. Abstract entities can still be explanatorily indispensable and explanatorily efficacious as well.

Pincock (2007) has pressed a point like this, but perhaps not far enough. His idea is that there are explanations and explanatory patterns (which he calls abstract or structural) which are part and parcel of the content of scientific theories. He restricts his attention to explanations that involve mathematical abstract objects (though it is clear he would not object to generalising them to NMAOs). These explanations proceed on the basis of descriptions of a physical system at a higher level of generality than its concrete physical constitution, by ignoring the microphysical properties of the system under study. (I would add: by ignoring the actual physical realisations of a NMAO, which in many cases might be inexact and approximate; that is, by replacing concrete physical systems by abstract physical systems, which are modelled/represented by a mathematical structure.) Pincock argues that this kind of explanation is important for the generality of explanations offered by physical theory. As noted above, the generality of physical explanation (and its applicability) requires abstract physical entities.

The conclusion Pincock draws is that mathematics is epistemically indispensable to science (because we are often ignorant of the detailed physical or microphysical facts that might realise a certain abstract explanatory pattern) and claims (2007: 263–4) that epistemic indispensability is compatible with metaphysical dispensability. To be sure, he goes on to qualify this point by noting that the metaphysical dispensability of mathematical objects cannot be warrantedly asserted because, strictly speaking, the nominalistic content of theories is indeterminate: 'our theories, understood in the light of the evidence we have, do not determine a collection of physical claims that we could view as the nominalistic content of these theories' (2007: 267). I do not doubt that Pincock is right, but his point is epistemic and as such it can only show that we have no good reasons, given the evidence, to take it for granted that mathematical entities are metaphysically dispensable. If what is said above is broadly correct, there is a stronger position to occupy, viz., the very idea of an abstract-entities-free nominalistic content of theories is hollow. Abstract entities (both non-mathematical and mathematical) get entry visas because very little interesting (general and explanatory) can be said about the physical world without being committed to them.

So far, I have claimed that the fact that abstract objects are causally inert is not a reason to make us suspicious about their ontic status. Still, the question remains: if they are not given to us causally, how are they given to us? The answer to this question is: via a (suitably generalised) Fregean context principle (CP). Briefly put, the idea is that terms that purport to refer to abstract objects have their reference fixed by the contribution these terms make to what is required to determine the truth of the sentences in which they occur. Reference, in other words, is fixed via the truth of sentences in which certain terms occur. Actually, CP is the only non-question-begging way to settle the issue of the reference of terms that purport to refer to abstract objects; hence, the only way to have access to abstract entities. Abstract objects can neither be encountered in experience nor be presented in it. They cannot be initiators, or links in, causal chains that end up in perceptual states. To think otherwise is to have the wrong idea of what abstract objects should be; it is to view abstract objects as actual physical objects stripped of their causal powers.[11] But abstract objects constitute a different kind of object. It is not surprising then that their mode of knowledge is different.[12]

Note that CP meshes very well with *IND-A*. The punch-line of the latter is that literal understanding fixes what conditions must be fulfilled for a theory to be true and truth determines that these conditions are satisfied: literal understanding + truth fix what is real. On this account, the real can be either concrete or abstract and there is no non-question begging further criterion to fix what should count as real. Unless explanation is conflated with causal explanation, abstract objects can be explanatorily relevant (say, by promoting unification), while not being causally relevant. And unless evidence is taken to be direct sensory evidence, there can be evidence for the reality of abstract objects via the evidence for the truth of theories in which they are being ineliminably referred.

5   On nominalistic adequacy

Let us *presume* we can make good sense of the idea of the nominalistic content of a theory and pay some attention to the concomitant idea of nominalistic adequacy. The significance of this idea is that a theory can be nominalistically adequate and yet false (in that there are no mathematical entities). It is further argued that a nominalistically adequate theory (which is not just an empir-

---

[11] Being causally active, that is having causal powers, is a criterion of objecthood. This is sometimes called the Eleatic Principle. According to the Eleatic Stranger in Plato's *Sophist* (247 D–E), '…everything which possesses any power of any kind, either to produce a change in anything or to be affected even in the least degree by the slightest cause, though it be only on one occasion, has real existence'. Graham Oddie (1982) has forcefully criticised this principle. In any case, it is not the only criterion of objecthood. Actually, making it a *sine qua non* condition for objecthood begs the question.

[12] For some stimulating discussion of the Fregean Context Principle in relation to nominalism, see Heck (2000).

ically adequate theory) is exactly as explanatory of the observable phenomena as a platonistically adequate theory that assumes the existence of abstract entities—since the latter make no contribution to the causal explanation of the observable.

There is first an issue with the very idea of characterising n-adequacy. Leng's characterisation, stated as it is in terms of the truth of nominalistically-stated consequences of a theory is problematic.[13]   As Jeff Ketland has noted (private communication), the required notion of n-adequacy should be model-theoretic. [14]

Briefly put, a theory T is n-adequate if a sub-structure of a model of the theory (this substructure which is fit for the representation of nominalistic facts) is isomorphic to the causal structure of the world. But now we have quantified over models—that is mathematical objects. Even if this objection is not fatal for the use of the concept of n-adequacy by the advocate of NSR, it would surely remove a lot of the attraction of NSR. Their advocates would have to have a fictionalist stance towards a central building block of their own account of how theories latch onto the world.

NSR forfeits the idea of a mathematics-free reformulation of scientific theories. This kind of situation leads to an interesting case of underdetermination, whereby the nominalistic content of a theory underdetermines its *full*

---

[13]Even if a syntactic characterisation of n-adequacy were adequate, the following would be a problem. Theories yield consequences only with the aid of auxiliary assumptions. The claim then of n-adequacy would have to be that a theory T is n-adequate if for *all* auxiliaries M cast in mathematical language, M&T yield no extra nominalist consequences that do not follow from T alone. If this were *not* the case, some of the n-content of T would depend on the truth of mathematical claims. The only way to ensure that this does not happen is to retreat to the conservativeness of mathematics.

[14]According to Ketland (private communication), to get to a characterisation of n-adequacy, we need:

> a) a 2-sorted theory-formulation language L, with two variable sorts (one ranging over concreta and the other ranging over abstracta), and three kinds of predicate: primary, mixed and secondary; and
> b) a partial nominalistic interpretation $I_N = (D_N, N_i)$ of L, where D_N is the domain of concreta and N_i are the nominalistic relations (which interpret the nominalistic predicates).

Let us call L_N the language without variables ranging over abstracta. Then:

> Def 1: An L-structure M is *nominalitically correct* iff its reduct to L_N is isomorphic to I_N.
> Def 2: A theory T is *weakly nominalistically adequate* iff all of its L_N-theorems are true in I_N.
> Def 3: A theory T is *nominalistically adequate* iff T has a model M whose L_N-reduct is isomorphic to I_N.

One can then show:

> Theorem 1. If T is n-adequate, then T is weakly n-adequate.
> Theorem 2. There are theories T that are *weakly* n-adequate, despite being n-*inadequate*.

Ketland has presented these ideas in his (2011), a paper delivered to the Aristotelian Society, 10 January 2011.

content. We can easily envisage a situation in which two (or more) theories T1 and T2 have exactly the same nominalistically-stated consequences but differ in their mathematical formulations. These theories are n-equivalent. To simplify matters let us assume that theories have two distinct and separate (or separable) parts, one nominalistic (call it N) and another mathematical (call it M). So a theory T is in effect N+M.

Suppose we take NSR to accept, as it surely should, the view that it is not a necessary truth that mathematical objects do *not* exist. Take, then, a theory T1 (= N+M) and another theory T2 (= N +(−M)). T2, in effect, asserts that there are no mathematical objects at all and equates the content of the theory with its n-content. T1 and T2 are n-equivalent. Yet, given that there could be mathematical objects, there is a possible world W1 in which there are mathematical entities and in this possible world T1 is true and not just n-adequate, while T2 is n-adequate but false. Similarly, there is a possible world W2, in which there are no mathematical objects, in which T2 is n-adequate *and* true. How can we tell whether the actual world @ is like W2 and not like W1? That is, how can we tell whether T2 is n-adequate and false as opposed to n-adequate and true? Given that we read the mathematical parts of our theories literally, as NSR agrees, @ could be like either W1 or W2 and, if anything, it is a contingent matter what it is like. The advocate of NSR simply lacks the resources to make all these distinctions and, in particular, to discriminate between all these worlds. It follows that NSR cannot simply assert that @ is like W2; nor can it assert that theories are n-adequate and false as opposed to n-adequate and true (in the sense that there are no mathematical entities). Unless there is an argument to the effect that necessarily, mathematical entities do not exist, the advocate of NSR can at best be an *agnostic* about their existence. Note that an appeal to Ockham's razor in this context would be question-begging. Given that n-adequacy underdetermines truth and that n-adequacy is all we have, the issue at stake is precisely to offer reasons to apply Ockham's razor to mathematical entities as opposed to remaining agnostic about the reality.

Here is another problem. Take T1 (= N+M) and T2 (= N+M′) such that they are n-equivalent. Since, according to NSR, there are no mathematical entities, T1 and T2 are both false. But there are two ways in which a theory can be false—one is when there are no mathematical entities and the other is when it asserts something false about a putative mathematical entity. So to say that 3 is composite is false on both counts, but to say that 3 is prime is false only on the first count. Envisage a situation in which T1 and T2 are such that a claim of the sort '3 is composite' is part of T2 and a claim of the sort '3 is prime' is part of T1.[15]

There is something deeply wrong with T2 but an advocate of NSR should tolerate it because it has no bearing on the nominalistic adequacy of T2 and its presumed n-equivalence with T1. A standard riposte by nominalists (when

---

[15]This example is, of course, merely illustrative of the general point. Other more serious examples can be easily found, e.g., related to non-Euclidean geometries.

a similar story is told about pure mathematics) is that '3 is prime' is true-in-the-story-of-mathematics, while '3 is composite' is false-in-the-story-of-mathematics. This kind of answer, whatever its merits in the case of pure mathematics, has *no* relevance to the present situation. If what matters is nominalistic adequacy and both theories are n-adequate, it is irrelevant that one of them is true-in-the-story-of-mathematics while the other is not, since, on the NSR view, truth-in-the-story-of-mathematics has *no* bearing on truth-in-the-story-of-physics, that is on n-adequacy. What follows from this problem is that there is a sense in which NSR cannot respect even the role of mathematics in science that NSR finds unobjectionable. Mathematics, of the standard variety, is not even theoretically and descriptively indispensable, since NSR cannot discriminate between false mathematical theories and those that are standardly used by mathematicians and physicists.

Here is yet another problem. Take T1 (= N+M) and T2 (= N+M') such that they are n-equivalent. For NSR, that's all that can be said of them. Any choice between them has no further epistemic relevance. But suppose that T1 (= N+M) is simpler, or more unified than T2(= N+M'). Suppose, that is, that the mathematical formulation M of T1 endows T1 with a number of theoretical virtues over the mathematical formulation M' of T2. For a scientific realist, theoretical virtues are truth-conducive. Hence, a scientific realist would have reasons to prefer T1 over T2 and to claim that T1 is more likely to be true than T2. But since T1 (= N+M) and 2 (= N+M') are n-equivalent, the respects in which T1 is more likely to be true than T2 should have to do with the mathematical content of T1, e.g., its abstract structural claims. Note that the kind of situation just envisaged cannot be circumvented by taking M and M' to be merely descriptive and representational devices. Any serious advocate of NSR would have to reformulate T1(= N+M) and 2 (= N+M') in such a way that they are mathematics-free, and *then show* that the reformulated T1 is simpler and more unified than the reformulated T2. There is no general reason to expect this to be the case. It will depend on the further axioms that are chosen and employed.[16]

A fully-fledged scientific theory is a theory proper. The claim that a theory T is n-adequate does not amount to presenting another theory. But let us accept, for the sake of the argument, that T_{na} is a theory: the nominalist-reduct of T. Take two theories T1 and T2 and conjoin them. T1&T2 will have extra non-nominalistic consequences and, in all probability, extra nominalistic ones. Put together, instead, T_{na1} and T_{na2}. For one, T_{na1}&T_{na2} might even be inconsistent (simply because they might have contradictory platonistic parts;

---

[16]Juha Saatsi (2007: 27) misses the point when he claims that 'no mathematical entity has ever been *introduced* as the best explanation of some (mathematical or physical) phenomena'. This claim seems oblivious to the facts that not all explanation in science is causal; that some explanations are unifications; and that some unifications—the most interesting ones—are effected only by introducing mathematical *structures*. It's irrelevant that these structures might not have themselves first been introduced 'as the best explanation of some phenomena' (2007: 27). How an entity is first introduced is independent of the explanatory work it does, after it has been introduced.

generally, T1 is n-adequate & T2 is n-adequate does not imply that T1&T2 is n-adequate.) But let us leave this to one side. Is there a guarantee that T_{na1}&T_{na2} has exactly all and only nominalistic consequences of T1&T2? The only way to secure this is via the conservativeness of mathematics. If mathematics is indeed conservative, then all and only n-consequences that follow from (T1&T2) + M will follow from T_{na1}&T_{na2}. There is nothing wrong with conservativeness *per se*. But the pertinent point is that any possible benefits from going for nominalistic adequacy of scientific theories instead of their truth requires the conservativeness of mathematics; the move to n-adequacy doesn't add much to whatever benefits already follow from conservativeness.[17] Leng (2005: 76-7) has stressed that commitment to n-adequacy provided an easier route to nominalism than Field's reliance on nominalisation-plus-the-conservativeness-of-mathematics. If I am right, Leng's claim is wrong.

One further notable reason to question the very idea of n-adequacy is that theories do not confront the phenomena (that is, *physical* occurrences) directly. Rather, they confront *models* of the data, which are mathematical structures (or more generally, NMAOs). The path of a planet, being an *ellipse*, is already a model of the data (and, in particular, a mathematical entity). The actual physical path is too messy to be of any use in the physical theory. Newton's theory accounts for the model of the data, that is the elliptic orbit (at least in the first instance). More generally, what really happens when a theory is applied to the world is that the theory is *first* applied to a model of the data, or to a heavily idealised (and mathematised) abstract physical object and *then* it is claimed that this model of the data captures some physical structure. As French (1999) has forcefully argued, inter-theoretic relations as well as relations between the theory and the world are ultimately, relations among mathematical structures.

Actually, this is not *very* surprising. Even if not all representation in science is based on isomorphism (or some other kind of morphism) a lot is so based and the very idea of structural similarity requires comparison between structures. But then, what is really compared when the n-adequacy of a theory is judged are two models, viz., two mathematical structures: the theoretical model and the model of the data. It is a further and separate claim that the model of the data (or the theoretical model for that matter) adequately represents concrete causal physical systems (or patterns). For the theory to be n-adequate it is the latter claim that has to be true. But this simply pushes

---

[17] In his (1996) James Hawthorne has proved that if scientific theories are properly fleshed out, then they will not contain excess non-mathematical content, and in particular that no excess non-mathematical (that is, nominalistic) content will be generated when they are conjoined with other such theories. But his proof holds under very special conditions, which require that there are representation theorems between a mathematical theory T1 and a non-mathematical theory T2 such that a) every sentence of T2 is a non-mathematical consequence of T1 and b) adding set theory to T2 yields all and only the consequences of the mathematical version of T1. As Hawthorne notes, this kind of proof cannot be general and does not follow from the conservativeness of set theory. That these representation theorems hold has to be proved individually for each and every pair of theories. It is obvious that this kind of strategy cannot be helpful to NSR.

the problem one step back. For now, the question is whether the model of the data itself (let's fix our attention on this to make things easier) is n-adequate vis-a-vis the appearances or the phenomena and answering this question pre-supposes either a direct confrontation of the model with the (unstructured) phenomena or the comparison of the model with another—that which (presumably) captures the *causal structure* of the phenomena. The first option does not seem to make much sense. The second option requires that the phenomena (or the world) has a built-in causal structure.

The key point here is not that this last assumption can be questioned.[18] Rather it is that the friends of NSR should come up with a conception of *the causal structure* of the world which is nominalist-friendly. This is too big an issue to be broached here. In the next section, I will simply try to motivate the thought that there is more in the world than just causal structure.

6   Mixed facts revisited

It was stressed quite early on, in Section 2, that marrying scientific realism with mathematical realism requires commitment to mixed facts—that is facts that are constituted by a combination of concrete and abstract objects. To fix our ideas, let us concentrate on Balaguer's (1998: 133) example. Take the mixed statement:

(A) The physical system S is at forty degrees Celsius.

Taken at face value, (A) expresses a *mixed* fact, viz., that the physical system S stands in the Celsius relation to the number 40. But the number 40 is causally inert. Hence, Balaguer argues, if (A) is true (as we presume it is), it is made true by two sets of facts that are independent of each other: a physical fact concerning the temperature of S and a platonistic fact involving the number 40. (A), Balaguer notes, does not express a *bottom-level* mixed fact; the truth that (A) expresses supervenes on more basic facts that are *not* mixed: a purely physical fact and a purely mathematical fact. But this suggests that (A) has a nominalistic content 'that captures its complete picture of S: that content is just that S holds its end of the '(A) bargain', that is, S does its part in making (A) true' (1998: 133). If this is the situation, Balaguer concludes, it can be the case that there are physical facts of the sort needed to make an empirical statement true, but no mathematical facts.

But is this the situation? Note, for a start, that what kind of facts make (A) true is not entirely clear. There is some physical fact, but in all probability, it has to do with the kinetic energy of S—a thing that, as Balaguer (1998: 133) adamantly admits, is not expressed by (A). So: the truthmaker of (A) is *some* physical state or other. Whatever that state is (if (A) is literally understood and

---

[18]For more on this, see van Fraassen (2006) and my (2006a).

true), it is such that it stands to a certain relation (the Celsius relation) to number 40. This number is, so to speak, the only non-negotiable part of the truth-maker of (A). What the physical state that stands in the Celsius relation to 40 is might vary (at least, we can be ignorant of it) but *that* it stands to this relation to *this* number is fixed (if (A) is true). It seems that the unity of the truth-maker of (A) requires this number, while it simply requires that there is *some or other* physical state related to this number, *modulo* the Celsius relation.[19] This unity is *not* a causal unity. It is not claimed that the mixed truthmaker is such that there is a *causal* relation C(S, 40) between the temperature-state of S (whatever that is) and 40. When it is asserted that the temperature-of-S-in-Celsius is 40 [(T_S-in-Celsius)=40] it is claimed that what is true of S concerning its temperature-in-Celsius is that it is equal to 40 and that this is true independently of what exactly it is that physically realises the temperature-state of S.

If we take '40' to be simply the name of a certain temperature state (we could have used the term 'Ralph' instead, as Balaguer notes in a different place), then all that (A) should be taken to express is that system S is in a certain temperature state which is designated by '40'. But then we implicitly admit that (A) has a purely nominalistic content and that it has no platonistic content. In other words, we do not read (A) literally; we do not honestly assume that it expresses a mixed fact (even if it is not a basic mixed fact). That S captures the complete picture that (A) paints of the physical world is simply causation-begging—it equates completeness with causal completeness.

Actually, the point just made generalises. In a lot of mixed statements, e.g., of the form 'Physical system S is (or forms, or constitutes) a group' (B), the abstract part of the truthmaker (e.g., the group) captures (again in a non-causal way) general structural features of the specific physical system as well as what a number of such systems (despite their diversity in physical terms) share in common. As Weyl put it, group theory reveals 'the essential features which are not contingent on a special form of the dynamical laws nor on special assumptions concerning the forces involved' (quoted by French 1999: 194).

Having said all this, I do not think that the facts that render statements like (A) are basic or bottom-level (though they are mixed). Clearly, that C(S, 40) is not bottom-level because it is a particular fact; and if we did admit it as bottom-level, we would have to make the massively uneconomical move to admit an infinity of bottom-level facts. Statements like (B) above are closer to being bottom-level, since they unify disparate physical systems. But in my view, bottom-level mixed facts are general facts that have to do with the structure of quantitative domains, the symmetry principles that characterise the world and things like that.

---

[19]This is a very important qualification, since I take to heart the Fregean lesson that the way numbers are attached to concrete objects is mediated by concepts (better: modes of presentation of these objects).

Balaguer is puzzled over the existence of mixed (physico-mathematical) facts. As has been noted, he takes it that mixed facts supervene on two independent sets of facts: physical *and* mathematical. Still, it does not follow that mathematical facts are dispensable *qua* parts of the truthmakers of mixed statements. Take the very simple statement 'The triangular road sign is yellow' (C). For this mixed statement to be true, a mixed fact is required, a part of which (so to speak) had to do with shapes and another with colours. There is no causal or other connection between shape-facts and colour-facts. We can even say that (C) has an S-content (whatever it asserts about shapes) and a C-content (whatever is asserts about colours). But though the two contents are independent of each other both are needed for the truth of (C), since even if each of them holds their end of the (C) bargain, none of them suffices to make (C) true.

7   A diagnosis and a conjecture

I have argued that NSR faces a number of problems in its attempt to motivate the weaker-than-full-truth notion of nominalistic adequacy. Even if we were to grant a clear and tolerably explained notion of n-adequacy, it would not follow that it would offer a better explanation of the success of science than the full truth of scientific theories. As noted already, discarding the abstract content of scientific theories (including the mathematical content) from being part of the best explanation of the success of theories is question-begging: it requires identifying explanation with causal explanation. The abstract content of theories plays a key role in ensuring the generality of the explanations offered and the unification of disparate phenomena in theoretical models. All this means that there is need for a more nuanced account of NMA (and of inference to the best explanation), where causal considerations are just one set out of many explanatory considerations. In my past writings (see my 1999, chapter 4) I too have put an emphasis on causal explanation. This has been wrong, especially insofar as it was meant to be exclusive of non-causal explanations. But clearly, not all explanation is causal—e.g., the explanation of low-level laws by reference to high-level ones. And explanation can also be of more abstract features of a system. Hence, even if causal explanation is indispensable, there is a more general level where the whole of the theory, with its abstract panoply, is seen as offering the best explanation.

There is a lot of hostility to abstract objects. Given their causal inertness, it is understandable—but, as I claimed, unjustified. But given this hostility, the friends of NSR take the view that the relevance of mathematics to empirical science has to do with our *understanding* and *representation* of the physical world and not with the operation of the physical world. If this is so, it can even be conceded that mathematical objects are theoretically and epistemologically indispensable, though, of course, metaphysically dispensable.

My own view is that Anti-Nominalistic Scientific Realism has to go all the way in its attack on NSR and claim that mathematical objects are part of the fabric of reality—though not in a way that has a causal impact on it. This implies that there are bottom-level mixed physico-mathematical facts. Actually, their existence seems to best explain the theoretical and epistemic indispensability of mathematics. But my conjecture is that though—to speak metaphorically—the concrete and the abstract co-operate to render mixed statements true, we are ignorant as to how this is done: as to what kind of unity a mixed physico-mathematical truthmaker has and in virtue of what it is united. More strongly, my conjecture is that we are cognitively closed to this kind of aspect of reality. I do not know how to back this up, but it seems to me it is the natural outcome of two predicaments: the first is that we tend to think of causation as the cement of the universe; the second is that, on reflection, we realise that the model of causal glue is too limited to account for the unity there is in the world—including the internal unity of the facts that make it up.[20]

[20]Versions of this paper have been presented in: the *PSA08* conference in Pittsburgh (November 2008); the *Metaphysics of Science* conference in the University of Melbourne (June 2009); and seminars in the Universities of Bristol (January 2009), Münster (April 2009), Barcelona (April 2010) and Milan–Bicocca (May 2010). Many thanks to audiences at these places for questions and comments—and especially to: Alexander Bird, Richard Boyd, Jim Brown, Mark Colyvan, Jose Diez, Brian Ellis, Geoff Hellman, Carl Hoefer, James Ladyman, Federico Laudisa, Mary Leng, Øystein Linnebo, Jose Martinez, Howard Sankey, Oliver Scholz, Christian Suhm and Nino Zanghi. Chris Pincock and Jeff Ketland deserve special thanks for their generous intellectual help and encouragement. A shorter version of this paper has appeared in *Philosophy of Science* (December 2010).

INDISPENSABILITY WITHOUT PLATONISM

Anne Newstead and James Franklin

1   Introduction

Indispensability arguments used to be the only game in town for philosophers of mathematics. One had to be realist about mathematics if one was a scientific realist. After all, mathematics is indispensable to formulating our best scientific theories. And it would be 'intellectually dishonest' to be realist about the physical components of scientific theory while remaining agnostic or anti-realist about the mathematical aspects of those theories.

Soon enough, however, the rot set in. Good philosophers began to have doubts about indispensability arguments. Parsons (1986) pointed out that the inferences to the best explanation mentioned in indispensability arguments didn't explain the 'obviousness' of elementary mathematical truths such as '2+2=4'. Furthermore, indispensability arguments leave unapplied pure mathematics in the twilight zone. In response, Quine dismissed such pure mathematics as 'recreational mathematics', surely a desperate move given that at any time a great deal of mathematics is unapplied.[1]

Then a strange thing happened. Even theorists in favour of the indispensability argument began to step back from embracing it wholeheartedly. Penelope Maddy led the way with her reminder that pure mathematics—such as set theory and analysis– is an autonomous discipline with its own distinctive epistemic practices and norms quite different from those employed in empirical

---

[1] For the debate over the significance of recreational mathematics, see Leng (2002) and the reply by Colyvan (2007).

sciences. In pure mathematics, mathematicians come to accept statements as true almost solely on the basis of *proof* from accepted axioms. It is surely an embarrassment for Quinean empiricism that it seems to get the epistemology of mathematics wrong. Its epistemological holism implied that mathematics should be tested and confirmed like the rest of empirical science. However, when an empirical theory fails to be confirmed, we don't take this failure as evidence that the mathematics used to articulate the theory is *false*. Rather, we don't even assume that the mathematics is being tested at all.[2]

It is bad enough that Quine's indispensability argument appears to distort the epistemology of mathematics. What is more scandalous is that philosophers cannot agree on the metaphysical conclusion of the argument. Sure, everyone agrees that indispensability is an argument for realism, but beyond this point the agreement ends. Quine's indispensability argument tells us nothing specific about the *metaphysical* nature of mathematical entities. It does not tell us *what* the basic mathematical entities are, or *in what way* they exist. It does not settle the ancient dispute between Platonists and Aristotelians over whether mathematical objects are abstract or concrete, particular or universal. The indispensability argument simply tells us that we ought to believe in the existence of whatever it is that mathematicians are talking about, because we are *ontologically committed* to them by our best scientific theories.

Despite brief protests to the contrary,[3] most scientific realists still assume that the conclusion of Quine's indispensability argument will involve some commitment to *abstract* entities.[4] In this assumption, realists are no doubt influenced by Quine's reluctant Platonism about classes at the end of *Word and Object* (1960: 233–70). Quine becomes a reluctant Platonist because he knows of no alternative way of construing classes and numbers other than as *abstract, other-worldly* entities. Deeper reflection on his indispensability argument shows that it is metaphysically shallow: the fact that such-and-such mathematics is useful in doing science tells us very little about the content of the metaphysics of science or mathematics. In fact, indispensability arguments are structurally as well as metaphysically neutral as regards the variety of realism we adopt: they don't tell us whether mathematical objects are abstract or concrete, lone atoms or structured complexes. (Similarly, arguments for the reality of atoms do not tell us whether they are hard particles or probability clouds.) Rather, indispensability arguments simply tell us that we ought to believe in the existence of mathematical objects, because we are ontologically committed to them by our best scientific theories.

---

[2] The point is made at length in Sober (1993).

[3] On attempts to make way for a non-Platonist variety of realism, see Cheyne and Pidgen (1996). The possibility is mentioned in passing in Colyvan (2001: 142).

[4] In his paper in this volume, Stathis Psillos argues that the indispensability argument does lead one to conclude that there are 'mixed facts', consisting of an abstract, mathematical component and concrete, physical component. On the Platonist realism that results, there need be no causal interaction between the abstract and physical components of such 'mixed facts'.

The search is now on to salvage what is left of indispensability arguments. The insight is that mathematics *works*: that in some sense mathematics must contain a body of truths because these truths can be exploited to describe and predict events in the world. And those truths are expressed in specifically mathematical language, mentioning functions, groups and other specifically mathematical entities. Mathematical explanations are successful, because (we infer) they correctly describe (the mathematical structure) of reality. Furthermore, this insight is strictly independent of Quinean philosophy. All it requires is application of the general argument for scientific realism (using inference to the best explanation) to the special case of mathematics. Arguably, this was Quine's intention originally. But in any case, a proper understanding of indispensability arguments must attempt to distance itself from its Quinean heritage. It is this act that we attempt in this essay: indispensability without Quineanism. In particular, we think that indispensability arguments for realism need not incorporate these dubious Quinean theses:

A. The Quinean criterion of ontological commitment: to be is merely to be the value of a bound variable in a canonical (first-order logic) statement of a theory.
B. Mathematics is no different epistemically from the rest of science.

In this essay we focus entirely on the task of liberating the indispensability argument from (A). The really unique aspect of our rejection of (A) is that we do so from a perspective that is not anti-realist, fictionalist, or nominalist, but from the perspective of *(neo-Aristotelian) realism*. A *realist* about a theory T is someone who (a) believes that T is true, and has determinate truth-values independently of whether we are in a position to verify those truth-values, and (b) believes that T describes some features of reality, and that therefore the features that T describes 'really exist'. For example, suppose T contains arithmetic. Then the realist believes that arithmetic has truths, that these truths are true anyway (independently of our coming to know them), and that the subject-matter of arithmetic 'really exists'.[5] Thus far (a) and (b) describe commitments that any realist shares. A *neo-Aristotelian realist* is someone who adds to commitments (a) and (b) some distinctive views about the nature of mathematical existence. Neo-Aristotelians hold that (c) basic mathematical patterns and universals are instantiated in nature (whether they can be exactly perceived or not), and that in the case of huge structures that may exceed what's found in nature, such structures *could be* instantiated even if they aren't (see Franklin 2009). David Armstrong's position on mathematical universals qualifies as neo-Aristotelian (Armstrong 1997; 2004: c.9). By contrast, Platonist

---

[5]Of course, it is a further matter to specify what the subject matter of arithmetic is. Some would say it is the structure of the natural numbers as described by the Dedekind and Peano axioms.

realists reject (c) on the grounds that mathematical universals are not perfectly instantiated in nature.[6]

For our part, we think many (perhaps not all) of the difficulties with the indispensability argument can be traced back to Quine's philosophy. His criterion is anti-Aristotelian because 'value of a bound variable'—especially with the emphasis on first-order logic—is intended to be read so that the values can only be *particulars*. Nominalism and Platonism share a commitment to the thesis that all entities are particulars (the Platonist admitting abstract particulars, the nominalist not). Aristotelianism denies that. We will clarify later in the paper Quine's allowing quantification to range over particulars but not properties.

Quine's criterion of ontological commitment—'to be is to be the value of a variable'—is part of the standard indispensability argument. We think Quine gets the ontology of mathematics wrong in several respects, all of which can be traced back to his application of his criterion of ontological commitment. First, Quine attempts to fit theories into the procrustean bed of first-order logic. Thus at a single stroke he excludes an ontological commitment to properties. Second, his criterion of ontological commitment is geared up to an atomist metaphysics, emphasizing individuals rather than states of affairs (facts), and complexes of individuals related to one another.

We propose an alternative to this atomist metaphysics, using what we might call Armstrong's new criterion of ontological commitment, 'to be is to be a truth-maker, or a component of a truthmaker'.[7] It is then possible to run a new indispensability argument with a different outcome. Of course, much depends again on what the truthmakers are. We follow Armstrong in supposing that the basic items in reality are *facts* as well as relations and properties. Arguably, this less atomistic and more relational approach is a better fit with the attractive view that mathematics is about patterns rather than objects. Whether one agrees with the resulting view or not, it demonstrates the possibility of a non-Quinean indispensability argument.

Section 1 below explains the involvement of Quine's criterion in traditional indispensability arguments. Section 2 puts forward Armstrong's alternative proposal for ontological commitment. It explains Armstrong's complaint that Quine is biased against properties in his criterion of ontological commitment. Section 3 presents a new indispensability argument that uses Armstrong's criterion of ontological commitment. Section 4 concludes that the new indispensability argument is better than the old one.

---

[6] On the issue of perfect versus imperfect instantiation, see Pettigrew (2009).

[7] Thanks to Jonathan Schaffer for the phrase. To call Armstrong's suggestion 'a criterion' is perhaps to sharpen it beyond what Armstrong had in mind. However, we let it stand for the sake of parity in discussing Quine and Armstrong on ontology.

2   The standard indispensability argument and its reliance on Quine's criterion of ontological commitment (OC)

We are concerned not so much with Quine exegesis as the indispensability argument as it has come to be known in wider philosophy of mathematics circles. Colyvan (2001: 11) provides a general outline of the key indispensability argument:

> (1) We ought to have ontological commitment to all and only the entities that are indispensable to our best scientific theories.
> (2) Mathematical entities are indispensable to our best scientific theories.
> (3) We ought to have ontological commitment to mathematical entities.

Ontological commitment figures twice in the argument, once in premise (1) and once in the conclusion (3). However, we are not told how to determine the ontological commitments of a theory. Colyvan refers to premise (1) as Quine's *ontic thesis* as opposed to Quine's actual thesis of *ontological commitment*. The idea is that (1) can serve as a general and normative premise about what considerations govern our ontological commitments without providing a recipe, 'a criterion', for ontological commitment. It is clear, though, that the Putnam–Quine version of the argument specifically invokes Quine's criterion of ontological commitment (OC). This is explicit in Putnam's version (1971: 57):

> So far I have been developing an argument for realism roughly along the following lines: quantification over mathematical entities is indispensable for science, both formal and physical: therefore we should accept such quantification; but this commits us to the existence of the mathematical entities in question. This type of argument stems of course from Quine, who has for years stressed the indispensability of quantification over mathematical entities and the intellectual dishonesty of denying the existence of what one daily presupposes.

We shall focus our discussion explicitly on this *quantificational form* of the indispensability argument. It may well be that there is a better form of the argument that is not so dependent on Quine's criterion of ontological commitment. Be that as it may, in this form of the argument, Quine's criterion of ontological commitment (OC) is used to explain the meaning of 'indispensability' in the original argument. The entities that are indispensable are just those that are in the domain quantified over by the canonical statement of our best theory.

In practice, however, we still know very little about our ontological commitments until we identify a specific theory and its language. Most theories in physics make use of functions on the real numbers and thus incorporate the mathematical theory of real analysis. The very notion of measurement involves mapping a quantitative property (heat, weight, mass, length, charge etc.) onto a real number. For example, we measure an inchworm and learn that it is approximately 3.5 cm. In practice, we can measure quantities by just rounding off decimals and reporting quantities as rational numbers. However, if we suppose that there are no gaps in our field of numbers and no limit to the exactness of measurement, we end up with something like the real number structure (as captured by the axioms of real analysis). The real number structure holds out the *ideal* of infinite precision.[8]

Moreover, it looks to be the case that real analysis (or some structural surrogate of it) cannot be dispensed with in our physics. If this is disputed, consider the fact that Field's attempt in *Science without Numbers* (1980) to eliminate reference to the real numbers from Newtonian mechanics simply ends up imposing the structure of the real numbers on a collection of spacetime points. Field finds this outcome acceptable as a nominalist because he urges that spacetime points are concrete entities, not abstract. But he admits he would not attempt to pursue physics finitistically. From a structuralist point of view, though, the real number structure *is* instantiated in Field's collection of spacetime points. That means that the real numbers have not really been eliminated from physics. Rather, we should think of the real numbers as a certain structure that exists physically (or could exist) rather than conceiving of them as the referents of linguistic terms that could be eliminated from the language of our scientific theory.[9]

So it is reasonable to suppose that Quine's criterion of ontological commitment applied to contemporary physics commits us to the existence of real numbers and functions on real numbers.[10] Thus, we can consider a more topic-specific version of the indispensability argument. Stewart Shapiro (2000: 228) presents one such version:

(1a) Real analysis refers to, and has variables range over, abstract objects called 'real numbers'. Moreover, one who accepts the truth

---

[8]Is it just that—an *ideal*? Maybe. It must be admitted that realism about the real numbers is harder than realism about rational numbers and natural numbers. One of the reasons for this is our measurements are never infinitely exact. For some considerations in favour of classical realism, see Newstead and Franklin (2008), and Newstead (2001).

[9]For criticism of Field on this point, see especially Resnik (1985).

[10]We ignore for a moment the real tension between the view of space-time as continuous that we find in Newtonian mechanics and GTR with the view in Quantum Mechanics that space-time is quantised. The natural way to interpret the real numbers physically is as points in a space-time manifold. QM raises doubts about whether we should preserve this physical interpretation of the real numbers. Schrödinger himself thought that the idea of a continuum was exposed by QM as a myth. It's fair to say that the jury is still out, but that Schrödinger's view has the most support among physicists.

of the axioms of real analysis is committed to the existence of these abstract entities.

(2a) Real analysis is indispensable for physics. That is, modern physics can be neither formulated nor practised without statements of real analysis.

(3a) If real analysis is indispensable for physics, then one who accepts physics as true of material reality is thereby committed to the truth of real analysis.

(4a) Physics is true, or nearly true.

The desired conclusion is:

(5a) Abstract entities called 'real numbers' exist.

Shapiro's version of the indispensability argument urges that in accepting physics as true, we are thereby ontologically committed to the real numbers. Of course, many of those who are unmoved by indispensability arguments don't really believe in the *truth*—in some heavy sense—of scientific theory in the first place. To be sure, one need not be committed to the *exact* truth of the laws of physics. The laws are idealizations which the physical phenomena approximate. Still, insofar as physical phenomena conform to the laws approximately, the laws are true 'nearly enough'.

If the truth of the scientific theory is accepted, then it becomes a straightforward matter to see why one would assume an ontological commitment in accepting the theory as true. On many substantive theories of truth, truths carry ontological commitments with them. For this very reason, some theorists view the indispensability argument as begging the question against fictionalism and instrumentalism. Savvy fictionalists (such as Leng 2005a) simply don't grant the substantive truth of scientific theories and explanations. This effectively blocks the inference to the reality of the items postulated by scientific theories. However, indispensability arguments target those who are *already* scientific realists, and thus would accept the truth ('near enough') of scientific theory. The point of the original indispensability argument was to show that scientific realists should not exempt mathematics from their realism.

Several other features of Shapiro's version of the argument deserve comment. Plainly, the abstractness of the entities in the conclusion is a result of the abstractness having been input in the first premises—by a sleight of hand, Shapiro builds into premise (1a) a conception of the real numbers as 'abstract entities', where presumably these real numbers are to be understood as non-spatiotemporal entities. This metaphysical conception of the real numbers is actually extraneous to the main argument. The vulgar conception of abstract objects is that they exist outside of space-time as Platonic universals. However, there is no need to hold a Platonist view about mathematical objects in order to maintain the indispensability argument. According to our view, known as 'neo-Aristotelian realism', we hold that universals are instantiated in nature, in our actual physical world. To be sure, if the physical universe is not infinitely large in extent or in the number of particles of a certain type that it contains, then some infinite structures will be universals that are merely possibly instantiated rather than concretely, actually instantiated in the physical universe. Even so it helps the epistemology of mathematics tremendously if we can count on there being a basic stock of mathematical universals that are exemplified in the world. For if a basic stock of mathematical universals is instantiated then basic knowledge of the universals can be gained through active perception and imagination.[11]

It is sometimes thought to be fatal to the Aristotelian philosophy of mathematics that certain mathematical forms (such as a square) are not visibly perfectly instantiated in nature. However, we note that there is a way around this problem. For example, it may be that although our perceptual experience does not always present a perfect square, our perceptual experience suffices to trigger in us the category specification of a perfect geometrical square. Thus, our perceptual experience can stimulate formation of exact mathematical concepts (Giaquinto 2007: 28; see also Newstead and Franklin 2010). Our perception is not fine-grained enough to allow us to discriminate between a perfect square and a very slightly imperfect square. The perception of a very slightly imperfect square is enough to induce in us the concept of a perfect square. This concept can then be used to form mathematical beliefs that are reliably related to perceptions of mathematical patterns.

There is thus no reason why a proponent of indispensability arguments for realism must accept, without arguments, the presuppositions of Platonist realism. Indeed, indispensability arguments are silent on the question of which variety of realism holds.[12] The metaphysical views that one extracts from indispensability arguments will be largely a function of the metaphysical views that one injects into such arguments. One primary place for the injection of metaphysics is in the specification of a criterion of ontological commitment; another place is in the selection of a canonical form for expressing the theory.

It is surprising, then, that Quine's criterion of ontological commitment has not been much criticized in the context of his indispensability argument. One recent exception is Azzouni (2004) who argues in favour of 'the separation thesis': we can accept scientific theories as true without being ontologically committed to the entities in the domain of quantification of the theory. Azzouni, therefore, rejects Quine's criterion and uses it to reject the indispensability argument. We also reject Quine's criterion of ontological commitment. We show, however, that we can recast the indispensability argument and perhaps inject new life into the argument by using a different approach to meta-ontology.

There is the starkest possible contrast between the separation thesis and the truthmaker approach to ontology. Truthmaker theorists believe that truth is inseparable from being to this extent: the truth of statements depends on

---

[11] For our position, see Franklin (2009).
[12] Indeed, even Platonists have agreed. See Colyvan (2001: 142).

being (on what there is in the world). If one accepts T as true, one is *ipso facto* committed to the existence of truthmakers for T.[13]

3   Armstrong's Alternative to Quine on Ontological Commitment

David Armstrong has given us two promising alternatives to Quine's criterion of ontological commitment. First, he has suggested that our criterion for the reality of an object obeys the *Eleatic Principle* (EP): everything that is real makes some causal difference to how the world is.[14] EP comes in handy in the battle against the Platonist's commitment to abstract objects, which are notoriously causally inert and thus (it seems, on most theories of knowledge) unknowable, mysterious, and inexplicable. However, there does appear to be difficulty in defending EP as a criterion of reality for some mathematical objects. The curvature of space-time is used to explain the behaviour of objects in general relativity, but the geometrical properties of space-time are not obviously causal powers.[15] Realists want to affirm the reality of these geometrical properties. Obviously in response to this kind of example it needs to be made clear that EP cannot simply be the slogan 'everything that exists is itself a causal power'. However, as we are inclined to adopt a neo-Aristotelian outlook in philosophy of mathematics in any case, we are glad to interpret EP in a different way than this simple slogan suggests. Neo-Aristotelians can hold that if EP holds true of mathematics, it has to do so in some way that acknowledges the difference between efficient causality and what we might call 'formal causality'. No one finds it plausible to say that mathematical quantities are efficiently causally efficacious, for example, in the same way that a billiard ball's motion of striking another billiard ball is efficiently causally efficacious. Nonetheless, perhaps mathematical quantities and patterns are causally implicated in the world in some other sense: they are part of a *formal* causal explanation of the world. For example, had the constants of nature been different, then objects in the world would behave differently. If G's value were different, then objects would not attract one another with the same gravitational force that they do. Although the notion of formal causality might seem opaque, it is at least strong enough to support counterfactual claims. Thus, if *x* is formally causally implicated in W, then the following counterfactual holds:

---

[13]There are a variety of views held by truthmaker theorists on the relation between truths and truthmakers. Various proposals for the relation include: supervenience, necessitation, and grounding. For a critical survey, see Schaffer (2008). Lewis (2001) advocated viewing the relation as supervenience, while Armstrong (2004) views the relation as one of necessitation.

[14]The *locus classicus* for EP are the remarks of the Eleatic stranger in Plato's *Sophist* 247e. Reference to EP in contemporary discussions originates with Oddie (1982). In Armstrong's work, see Armstrong (1997: 41) and for the 'truthmaker version', see Armstrong (2004: 7), 'every truthmaker should make some contribution to the causal order of the actual world'.

[15]See Colyvan (2001: c.3) for objections to EP.

had x not obtained, then some event e in W would not obtain either. Mathematical quantities are clearly formally causally operative in a counter-factually sustaining way.[16] For example, if the peg had not been square, it would have fit in the round hole. If we peer this far down the road to interpreting EP, we see that the disagreement between Platonists and others over EP gives way to a debate about how to understand causal-explanatory relations.

To be sure, one can both accept Quine's criterion of ontological commitment and EP, but in practice it seems better to have EP supplant Quine's criterion with EP altogether. The tendency of Quine's criterion is to allow into our ontology every individual over which our theories range, whereas the tendency of EP is to restrict our ontological commitment to some smaller class of entities that are the real players in our theory. We cannot enter into this debate fully here, but record it as yet another approach to doing ontology that provides a distinct realist alternative to Quine's criterion of ontological commitment.

The second alternative to Quine's criterion of ontological commitment derives from the theory of truthmaking.[17] According to the theory of truthmaking, every truth has a truthmaker, where this truthmaker is some entity in the world in virtue of which the truth is true. On Armstrong's particular metaphysics, it is indeed the case that *every* truth has a truthmaker (truthmaker maximalism), and further the case that the main truthmakers are *facts* or *states of affairs*. The key intuition is that truth is grounded in reality. In the absence of truthmakers for a given truth, the truth would 'float free' of how the world is. Such 'free floating' truths strike truthmaker theorists as unacceptable.

The truthmaker approach to metaphysics is certainly appealing to realists, but doesn't suppose a particular form of realist metaphysics.[18] Someone with a basic ontology of things (rather than facts) could allow that X was a truthmaker for each truth of the form 'X exists', where X names some concrete particular (as Armstrong notes, 2004: 24). In such a world of things, the fundamental truths would all have the form 'X exists'.

Nonetheless, it is of course true that the truthmaker principle does exact some commitment to realism about the truth-values of propositions/statements. The truthmaker theory does assume a kind of bland, minimal realism about truthmakers. Truthmaker theory states that for every (basic) truth, there is some truthmaker in the world. As these truthmakers enjoy a mind-independent existence, it follows that truthmaker theory is realist about the existence of truthmakers. The key point, however, is that truthmaker

---

[16] For an outline of how to pursue such an approach, readers might consult Bigelow and Pargetter (1990: c.8). Recently, Aidan Lyon (forthcoming) suggests that mathematical items are part of a 'programming' explanation of how things work; that is, part of the high level description that explains why we see the transitions between given inputs and outputs that we see. This suggestion may be a more contemporary way of phrasing the Aristotelian claim about mathematical patterns and quantities being formally causally explanatory of the world.

[17] See Armstrong (2004) for a basic exposition.

[18] We have been helped by reading Cameron (2008).

theory does not identify the truthmakers for us. There is no automatic way to move from a statement to identification of the truthmaker for that statement. In particular, no amount of analysis of the logical form of a statement—without doing some serious metaphysics—is going to tell us what the truthmakers are. Russell's logical atomism made this mistake, and Armstrong (2004: 23) does not repeat it.

> The truthmaker *method* suggests, then, a very general way of doing metaphysics:

> 'To postulate certain truthmakers for certain truths is to admit those truthmakers into one's ontology. The complete range of truthmakers admitted constitutes a metaphysics ... '

Armstrong emphasizes that the hunt for truthmakers is as hard an enterprise as doing metaphysics itself or science. Our ontological commitments will depend on our having identified a true theory of nature. Given a disdain for purely armchair science and metaphysics, this theory of nature will be determined *a posteriori*. For example, if it should turn out that everything is made out of sub-atomic particles such as quarks and gluons, then perhaps the truthmakers for certain statements about the physical world such as 'There's a table' will be complex facts about how sub-atomic particles are arranged in a certain space. That means to a certain extent that the contemporary metaphysician must wait on science. According to Armstrong's *a posteriori realism*, science will discover and identify the basic universals.[19] At best, the metaphysician can hazard a guess about the general structure of the truthmakers that will satisfy our best scientific theories.

To remain faithful to his *a posteriori realism*, Armstrong warns that truthmaker theory is only 'a promising way to regiment metaphysics ... not a royal road' (Armstrong 2004: 22). Nonetheless, it is tempting to harden his theory into a criterion for ontological commitment. The slogan for ontological commitment on Armstrong's theory is therefore 'to be is to be a truthmaker (or part of one) for a true theory'.[20] We have borrowed this slogan from Schaffer (2009) and amended it by adding 'or part of one'.

How will our ontological commitments differ from those of a Quinean, supposing that both followers of Armstrong and Quine are assessing the same scientific theory? In particular, how will our mathematical ontology differ? We contend that following Armstrong's 'truthmaker' approach will result in a richer mathematical ontology that includes *properties, relations,* and *facts*.

Consider the statements:

---

[19]The term 'a posteriori realism' is used by Mumford (2007) to describe Armstrong's position.

[20]Adapted from Schaffer (2009). We would prefer 'to be is to be the value of a truthmaker or *one of its components* thereof'. Consider the statement 'This square is red'. The property of being *red* is one of the components of the fact (this square's being red) that makes the statement true. On Armstrong's view, the main metaphysical commitment is to the fact or state of affairs of *this-square's-being-red*. However, the primacy of facts doesn't undermine the real existence of its components, *this square* and the property of *being red* (which is partly instantiated in this square)).

(1) Fa

(2) ∃x (Fx)

Quine thinks (2) makes plain the ontological commitment of the simple statement (1). If one accepts 'Fa' as true, then one's ontological commitment amounts to this: *there is something* that is F. One's ontological commitment is to some particular with some property called 'F'—but not to some property F instantiated in some particular a: one need not view 'F' as naming a universal property, and one need not adopt a realist view of properties. If one likes one can read (2) in a functionalist manner as saying that *there is something that plays the role of being F*. If one were further committed to the reality of roles (on the grounds of the theory's being 'heavily' true in some realist way), then re-iterating the Quinean procedure would suggest one also accept:

(3) ∃x ∃F (Fx)

In (3) the commitment to the existence of an object and a property is made explicit. But Quineans do not think that (1) and (2) imply (3), because one might accept (1) or (2) as true, without being committed to the separate 'existence' (as asserted by the existential predicate) of F. This raises the spectre that one might accept the truth of a statement 'a is F' while being deflationary in metaphysical terms about what this truth requires. Fiction is one area where we are used to this phenomenon. For example, 'Santa Claus has a beard' is true at least in the context of the Santa Claus story, but there is no bearded individual in the world that makes this statement true. However, in lieu of an argument for treating the statements of our scientific theories as fiction, the Quinean needs good reasons to block the move from (2) to (3). It seems that only a bias against second-order logic blocks the move. The bias against second-order logic, though, is mainly grounded in a distrust of properties as obscure entities lacking clear-cut individuation criteria.

Aristotelian realists such as Armstrong and his defenders argue that one needs the property F, the particular a, and also the fact of a's being F, to exist in order to make (1) true. According to (1), there is some particular that is F. This *something* cannot be a bare particular; it must have properties too. If 'Fa' is true, then there is *something* that has the property called 'F'. In accepting (1) one is committed to there being *something* (called 'a') possessing *some property* (called 'F').[21]

Armstrong for his part has long viewed Quine as guilty of 'ostrich nominalism': Quine thinks he can accept the truth of a statements such as 'a is F' ('That house is red') and 'b is F' ('That sunset is red') but not incur any ontological commitment to the property of being F (red) (Armstrong 1978: 16).

---

[21] But why stop here? The particular a and the property F must be related somehow, since 'a is F' asserts that a has F-ness, not just the existence of a and F unrelated. Armstrong proposes we take the state of affairs (or fact) of a's being F as the truthmaker for 'a is F'. One may also point out that one is committed to the components of the fact of a's being F which are the individual a and the property F, since facts supervene on their components.

Quine refrains from analysing '*a* is F' in such a way that it implies that there is a property of F-ness. However, in doing so Quine is left without the resources to explain in what respect individuals *a* and *b* resemble each other (as regards colour). The ancient 'one over many' argument posits universals (shared properties) as an answer to such puzzles. There are thus legitimate arguments for universals that go unanswered by Quine (e.g. Armstrong 1978: c.6; Armstrong 1997: c.3.). It will not do simply to dismiss the reality of universals (properties) by logical fiat.

As one might expect, Quine's analysis of the truth '*a* is F' offers a desert landscape: an ontological commitment to the lone individual called '*a*' that might satisfy the open sentence '_ is F'. Quine lacks the knowledge of Australians that deserts are not barren, but teeming with life. Armstrong's Australian picture of the matter is a dense, fertile landscape. The metaphysics required for the truth of '*a* is F' include an object *a*, its property *F*, and *the fact that a is F*.

4   The Indispensability Argument Revised

How now does the indispensability argument look if we run it using Armstrong's approach to ontological commitment? As we saw in the previous section, Armstrong's approach contains several components:

> (a) Truthmaker theory (which includes at a minimum the claim that every truth has a truthmaker together with some account of the truth-making relation).
> (b) Armstrong's own particular metaphysics, which identifies facts (states of affairs) as the main truthmakers, allowing for components of those facts (properties, relations, objects) as real existents.

We are going to apply (a) and a rather loose interpretation of (b) to the indispensability argument we considered earlier. In doing so—as is typical of the approach to metaphysics by hunting down truthmakers—we have to identify the particular truthmakers for a set of truths by examining those truths themselves and the practice in which they are found.

The old indispensability argument (1a–5a) claims that the truths of real analysis are indispensable to physics. We think the argument is correct in finding real analysis to be indispensable for physics. So, assuming that real analysis is indispensable to physics, we need to identify the truthmakers of real analysis. It is here that we go beyond truthmaker theory to offer a particular metaphysical claim about the nature of mathematical truthmakers. Our speculation is in keeping with Armstrong's metaphysics, although it is not specifically his view. Our view is that one of the main truthmakers for real analysis is the standard real number structure as found in any real number continuum. It is this structure which is described by the axioms of real analysis. These axioms include claims such as:

> (Btw) Between any two real numbers x and y, there is another real number z.
> (UB) Any set of real numbers with an upper bound has a least upper bound.
> (Archimedean Property) For any positive numbers x and y where $x < y$, there is some natural number n such that $nx > y$.

In addition to the continuum, real analysis also makes claims about functions and their properties such as differentiability, continuity, and integrability. So perhaps these properties should be taken as components of the facts that are the truthmakers for classical real analysis.

How does the indispensability argument look if we run it using Armstrong's criterion of ontological commitment? Remember that since Armstrong's 'truthmaker' criterion of ontological commitment is formal, we will need to supplement it with our preferred identification of the truthmakers of analysis. Here's how the revised argument looks:

> (1) The statements of real analysis concern truths about the real number continuum, both its subsets (sequences of the real numbers), the properties of those subsets (e.g. convergence) and all the functions that can be defined on subsets of the real number continuum, along with the properties of those functions (e.g. differentiability, smoothness etc.).
> (2) The truthmakers for statements in real analysis include sequences of real numbers and functions with the relevant properties. One who accepts the truths of the axioms of real analysis is committed to the existence of these mathematical entities. (Note that as usual reference to the real numbers is not to abstract entities called 'the real numbers' but to a structure, the real number continuum, that could be realised in space.)

The rest of the argument is unchanged:

> (3) Real analysis is indispensable for physics. That is, modern physics can be neither formulated nor practised without statements of real analysis.
> (4) If real analysis is indispensable for physics, then one who accepts physics as true of material reality is thereby committed to the truth of real analysis.
> (5) Physics is true, or nearly true.

The immediate conclusion of the argument is that we are committed to the existence of the truthmakers of real analysis. These truthmakers have been identified in step (2) of the argument as the sequences and functions of real numbers with the properties studied in real analysis (such as convergence, differentiability etc.). So the final conclusion is:

> (6) We are committed to the truthmakers of real analysis. These include (perhaps) the real number structure, real-valued functions, and the properties of real numbers and real-valued functions.

We stress that the conclusion is contingent on our having the correct identification of the truthmakers of real analysis. Moreover, identification of such truthmakers is a matter for those thinking about the metaphysics of mathematics. In doing so, one should bear in mind how the mathematics is being applied. However, we cannot expect an indispensability argument to tell us straight out what those truthmakers are.

5   One Problem with the New Indispensability Argument

We need to deal with problems that arise for our version of the indispensability argument, specifically from the fact that the mathematical theory under scrutiny is *real analysis*. While no one doubts that the ontology of classical real analysis includes an uncountably infinite real-number continuum, there are legitimate questions about the relation of the mathematical continuum to the structure of space-time. Whether space-time has a continuum-structure or a grainy structure is an empirical question. Thus far the evidence is equivocal, but leans towards suggesting the structure of space-time is grainy and not continuous (Wolfram 2002).

There are two possible solutions. Aristotle's own solution was to hold that the points of a continuum do not actually exist all at once. Rather, a point comes into being when we undertake an activity, such as dividing a line. Prior to such activity on the part of the mathematician, the point exists only potentially as the boundary of a line segment. The upshot of this view is that the truthmaker for many statements of real analysis could be a merely *possible* mathematical continuum. There is no need to be wedded to the view that there is a (physical) continuum in space-time.

Another possible solution is to revise our notion of which part of mathematics is indispensable for physics. Maybe real analysis is not indispensable, but some weaker form of real analysis is. Perhaps an exact mathematical description of the physical universe does not involve real analysis with its commitment to infinite divisibility. Instead the appropriate mathematics would be discrete analysis in which, for example, limits as Δx tends to 0 are replaced by ersatz limits as Δx tends to $\hbar$ (the size of an atom of space or time). Discrete analysis is mathematically legitimate, however, cumbersome (Zeilberger 2004). The main philosophical point, however, is that its ontological commitments are to the same kind of entities as real analysis: (discrete) functions which possess properties such as ersatz convergence and differentiability. The indispensability argument goes through with these entities and properties rather than the conventional ones. A kind of mathematical realism is still vindicated.

6   Conclusion

It is clear that running the indispensability argument with Armstrong's total approach to ontology results in a *qualitatively* richer ontology than the one offered by Quine. The mathematics that proves indispensable includes not just sets, but mathematical properties and facts about these properties and relations. But Quine's mathematical ontology is *quantitatively richer*: it allows unlimited numbers of classes. As Aristotelian realists, we would prefer to posit no more structures than we absolutely need to do the applied science: the rest might be uninstantiated structures of the sort posited by Platonism. We still think it's a gain to have one's basic structures be natural structures, however. In this way knowledge of such structures becomes less mysterious than knowledge of Platonic forms.

Our modest aim has been to delineate a possible position in logical space: realism about mathematics without Platonism, but motivated (in part) by indispensability considerations. We have shown that indispensability arguments can be run free of Quinean ontological baggage, such as Quine's criterion of ontological commitment. In its place we have suggested that the truthmaker approach to ontology might be preferable. We have tried to explain what such a view might look like, although in completing this task we needed to come up with our own preferred metaphysics of mathematics: Aristotelian realism (Franklin 2009).

We now pause to consider the peculiarity of our procedure. We have invoked truthmaker theory in our indispensability argument. But the indispensability argument is supposed to be an argument for realism on independent grounds—it shouldn't assume realism about mathematics. Doesn't insisting that the truths of mathematics have truthmakers assume realism about mathematics? We answer that it does assume semantic value realism (the truths of mathematics—guess what?!—have truth-values) but it does not assume a particular form of metaphysical realism. Truthmaker theory is itself agnostic about the identity of truthmakers for a particular theory, such as real analysis in mathematics. We have our favourite view of the existence of these truthmakers as Aristotelian realists. But our Aristotelian realism is a commitment beyond truthmaker theory, and not one that we expect everyone to share. Given our modest aim of establishing the viability of an alternative to Quine's Platonist indispensability argument, it would still be consistent with the letter of our position if all indispensability arguments were to be shown to reach the conclusion of realism by assuming realism at the outset. We don't think this would be a desirable outcome, but it is a possibility. Valid arguments can be question-begging, of course. To avoid begging the question we would want to have reasons independent of realism about mathematics for thinking that truthmaking was a good approach to determining the ontology of science.

We think that indispensability arguments provide compelling reasons to be realist, but not to be Platonist. The standard Quine-Putnam version of the argument relies on Quine's quantificational criterion of ontological commitment. It also imports a specifically Platonist version of realism in its suggestion that numbers and sets are 'abstract objects' (conceived of as existing outside of space and time). These metaphysical biases are not essential to the indispensability argument.

We suggest that another version of indispensability is preferable. We have suggested that we replace Quine's criterion with Armstrong's truthmaker criterion: 'to be is to be a truthmaker, or part of one, for a true theory'. We then tried to apply Armstrong's truthmaker approach to determine the ontological commitments of mathematical theories taking the theory of real analysis as our case study. We suggested that application of truthmaker suggests a mathematical ontology in which the fundamental items of mathematics are not lone objects, but patterns, properties, functions, facts, and relations. Such a qualitatively multifarious ontology—an Armstrongian bush, not a Quinean desert—might have advantages when it comes to maintaining a naturalistic epistemology.

Part IV

DISPOSITIONS AND CAUSAL POWERS

CAUSAL DISPOSITIONALISM

Stephen Mumford and Rani Lill Anjum

1   Causation for Dispositionalists

There are various ways that a dispositionalist could go when constructing a theory of causation based on an ontology of real dispositions or what some prefer to call powers. In this paper, we will try to spell out what we take to be the most promising version of causal dispositionalism. The broad aim is to get causes from powers. Many people share this aim and, as Molnar has already said, one of the reasons to accept an ontology of powers is the work powers can do in explaining a host of other problems, causation being among them (Molnar 2003: 186). The delivery of a plausible powers-based theory of causation is, however, overdue. So far we have only hints and false starts (Harré and Madden 1975, Bhaskar 1975, Cartwright 1989, Molnar 2003: ch. 12, Martin 2008: ch. 5). None of these accounts have gone quite in the right direction, in our view, so our ambition here is to set out the first few steps.

Let us begin by assuming the reality of dispositions or powers. There is much debate on the general issue of dispositions (Armstrong et al. 1996, Mumford 1998, Ellis 2001, Molnar 2003, Bird 2007a); but if we are to make any progress on the issue of causation, we had better leave the former debate alone. We are taking it for granted that in the world there are powers that naturally dispose towards certain outcomes or manifestations. These powers exist in objects and substances and are as real and objective as any other of those things' qualities, such as their height, shape or mass. Although such powers can issue in manifestations, they nevertheless can exist without them and may never manifest at all. Importantly, we reject the proposal that disposition ascriptions can truly be reducible to conditional statements. If that were true, there would be no need to be realist about dispositions at all. If the conditionals used only occurrent or categorical terms then dispositions would be reducible. We think, however, there is a general and systematic reason why the conditional analysis of dispositions could never work, and we will mention this shortly. This would explain why counterexamples to all the proposed conditional analyses have not been hard to come by.

So far, this has all been mandatory in order to get started and enough might already have been said to show that there ought to be some connection between real dispositions and causation. We have stated that powers can issue in certain outcomes and this of course resembles the notion of cause and effect with which we all grapple. There is still plenty of work for us to do, however, in explaining how we get causes from powers.

Before getting on to that detail, however, there is one non-mandatory step that we should signal we are taking. There is more than one general ontology that admits the reality of dispositions. Categoricalism doesn't, but both property dualism and pandispositionalism do. In property dualism, some properties are essentially dispositional but some are not. Our own sympathies are with pandispositionalism. We will be offering a view of causation based on all properties being powerful. Having some properties that are not, or at least are not essentially so, rather complicates the matter. Something would have to be said about these other, essentially powerless, properties and how they deserve their place in our ontology. We would also have to explain how the powerful and non-powerful properties related or interacted, if at all, which also seems no easy task. Instead, we go with the view that all properties are powerful and thus with pandispositionalism (see Mumford 2008).

Assuming pandispositionalism, every power's manifestation is also a power for some further manifestation. Armstrong (2005: 314) sees that this could make of causation the mere passing around of powers. He thinks that is a bad thing, while we think it gives a very good image of causation (Mumford 2009). To take Hume's billiard table, which he claims provides causation's 'perfect instance' (1740: 137)), the various balls crashing into each other are thereby passing on power. The first ball rolls across the table, its momentum being a power to move. When it strikes the object ball, it passes on that power. Momentum transfers from cue ball to object ball. And now that the object ball has that momentum, it too can pass it on to any other ball with which it collides. It need not always be the same power that is passed on, however. Causation often involves change but the change need not just be in the cause passing on the same property to its effect, as in the momentum of the billiard balls. Dropping a fragile vase on to a hard surface results in it breaking, which is as good a case of causation as any. But the vase has powers when broken that it did not have before: its pieces can now cut. And its power to hold water is lost when it becomes broken. Powers are passed on to the manifestation, but different powers from our original disposition.

There is a further question that needs to be addressed of the relation between singular and general causal truths. What is the dispositionalist's account of this? For general causal truths, we interpret them as one type disposing to-wards another type, for instance, such as that smoking causes cancer. It is important to read this dispositionally. Smoking only disposes towards getting cancer. We all know that some who smoke do not get cancer but the general causal claim can remain both useful and true if we read it dispositionally. If general causal truths were interpreted as universally quantified conditionals, then it would have to come out as false that smoking causes cancer since not everyone who smokes develops cancer. But if our general causal claims were restricted only to exceptionless regularities, then we would miss most, if not all, of the general causal facts. The dispositionalist solution is to say that there is an irreducibly dispositional connection involved between smoking and can-cer. The connection is less than necessary, because clearly there are some smokers who do not get cancer. But the connection is more than purely con-tingent. It is no mere coincidence that many who smoke do get cancer. Typ-ically, the chance of some cancers will be far greater for smokers than non-smokers. The dispositions of things would be the chance-raisers: the worldly truthmakers of the chancy truths.

For singular causal truths, we say that they concern one particular dispos-ing towards another, such as the striking of this match disposing towards it lighting. Most typically, though not necessarily always, singular causal claims often contain a success element. If I say (past tense) that my uncle's smok-ing caused his cancer, I am saying not just that his smoking disposed towards cancer but that it also 'succeeded' in manifesting its disposition. Saying that token a caused token b entails not just the more-than-contingent less-than-necessary dispositional connection between a and b, but also that b occurred and that b was the manifestation of a's disposition. We can say the same about most future and present tense singular causal claims. In contrast, general causal claims do not entail the further commitments but only that there is a dispositional connection between types A and B. There may well be some to-kens of those types that do not manifest their disposition and it is even a pos-sibility, if a remote one, that none of the tokens do.

Returning to the three elements of the typical singular causal truth, we can see that all are required. Only the third element might be challenged, as it could be ventured that the causal claim only amounts to a disposing towards b and b indeed occurring. But this is inadequate for dispositionalism. There might have been some other disposition c that also disposed towards b and it, rather than a, might have manifested b. Where we say that a caused b, we must mean that a specifically did at least some of the causing of b. Perhaps some events are uncaused altogether, so a true causal claim must be about more than just something being disposed towards and occurring. A causal claim is one of responsibility.

2   Contrasts

The powers ontology has its roots in Aristotelianism in which nature is active. The chief contrast is with Humean views of the world as a succession of events or facts in which we see patterns and project our future expectations on to the world. Setting aside Hume himself, the main contemporary advocate of this view is Lewis, who developed the influential view that causation consists in (the ancestral of) a counterfactual dependence between events (Lewis 1973a). Offering an account of counterfactuals within an essentially Humean, extensionalist framework is one of his great achievements (Lewis 1973b), though few find it easy to believe that there really is a plurality of other concrete worlds, more or less like ours.

The first major contrast between our account and a Humean and Lewisian one is that they offer a reductive analysis of causation, which we do not attempt. The notion of causation, in their view, can be cashed out in non-causal terms. For Hume, causation is understood just as a constant conjunction of event types, in which each cause is also spatially contiguous and temporally prior to its effect. For Lewis, causation is understood as a counterfactual dependence between events, where that amounts merely to the fact(s) that the cause and effect both occur: but in all the closest possible worlds in which the cause does not occur, the effect does not occur either. We have doubts that these, or any other reductive analysis, will be a success. Both are vulnerable to counterexamples: constant conjunctions without causation, causation without constant conjunctions, counterfactual dependences between events without causation and causation without counterfactual dependence. We will not go into the details of these counterexamples at this point, though some will be discussed later that relate to Lewis's account.

Not every concept permits analysis into others. Some concepts may be learnt directly from experience. This does not have to mean that the concept of causation is entirely simple. We think it has two parts, as we will explain shortly, but each part has to be experienced. Causation is one of our first experiences, both as a patient and agent, and it is one of our most basic, fundamental and important. Through our bodies, we act and are acted upon and have an understanding of this as soon as we have an understanding.

According to causal dispositionalism, causation involves an irreducible dispositional modality.[1] It is about one thing tending towards another, rather than necessitating it or the two being contingent accidents. If this is right, then we can never say that if some condition C occurs, no matter how large or complex C is, then an effect E will occur. We can only say that it is disposed to happen, will tend to do so, or that it is more or less likely to happen. These are ways of gesturing towards the dispositional modality but they cannot analyse it. They offer synonyms for it, and thus cannot be used in a reductive analysis. But the alternative is that non-dispositional terms would be used that failed to

---

[1] For more on the dispositional modality, see Mumford and Anjum 2011.

capture accurately the full notion of dispositionality. The putative conditional analysis of dispositions is one that fails for this reason. Something disposing towards something else can never be captured by an analysis of the form *if S, in conditions C, then M*. Even if a disposition is stimulated, and in the right sort of conditions to manifest, it still only tends or disposes towards its manifestation. It doesn't guarantee it. This is why we have a rich, specifically dispositional vocabulary available to us. We cannot replace it, without loss, by the idea of a conditional where, if the antecedent is true, and any associated conditions, the consequent *must* also be true. That has some resemblance to dispositionality, but is not close enough. We only want that if the antecedent is true, the consequent tends to be true, and this only restates dispositionality rather than analyses it.

How, then can the idea of a dispositional modality be acquired through experience? There must be two such kinds of experience, in our view. In the case where we are causal patients, we must first experience some power acting on our bodies, such as when we walk in a gale which could blow us over. But, second, we feel that we are able to resist it. The gale is for some manifestation, but it can be prevented. Where we are causal agents, it is us trying to manifest a power, such as the power to pull something. But, again, we sometimes feel that it can be resisted, such as when something else pulls in the opposite direction. These two components give us the idea of a power being for some specific manifestation to which its relation is more than contingent, but also capable of being prevented and hence to which its relation is less than necessary.

While we do not see how this dispositional modality can be explained non-circularly in non-dispositional, non-modal terms, this is not to say that it is mysterious or little understood. Indeed, we think it is the most familiar modality to any causally engaged experiencer, which virtually all humans are. And given that causation, on our account, essentially involves this, then we have to be primitivists about causation, for it contains an unanalysable element.

So what is the point of a dispositional theory of causation if it does not give us an analysis? There are still reasons to be interested in a theory that falls short of analysis. One is that, if what we have said is right, the theory tells us why an analysis is neither possible nor required, which is an important finding itself. Second, however, the dispositional theory emphasises causation's dispositional nature, which many following in Hume's wake have overlooked. The Humean tradition has concentrated on constant conjunction, and many post-Humeans have thought that the way to improve his theory was to have something in addition, such that constant conjunction has been judged a necessary but not sufficient condition for causation. The dispositional account, in contrast, tells us that constant conjunction is not even a necessary condition for causation. Indeed, where two phenomena really are constantly conjoined, the dispositional view tells us that there is not a causal connection between them but something else, such as identity or a truth of essentialism. Finally, while the notion of disposition is clearly itself a causal notion, there are also dispositional accounts of properties, laws, modality, and other things. The theory therefore has a unificatory potential.

3   What is missed by the counterfactual dependence theory but not by dispositionalism

One way in which to see the power of the dispositional view is to compare and contrast it with another of the leading theories. We choose Lewis's counterfactual dependence theory. There is a major difference at the outset. Lewis takes 'a cause to be only one indispensable part, not the whole, of the total situation that is followed by the effect in accordance with a law' (1973a: 159). We are not going to object again to the idea of the cause being followed by an effect in accordance with a law, which we take to be Lewis tipping his cap to constant conjunction. What instead we now want to draw to the attention is that Lewis is interested primarily in what it is to be *a* cause, rather than *the* cause. Dispositionalism has something to say about both and indeed shows how some important features of causation are missed by neglecting the latter.

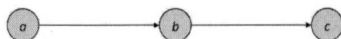

Figure 1: Standard neuron diagram showing stimulatory connections

We can see this when we consider the inadequacies of neuron diagrams, which have become the standard way to represent causal situations since Lewis's work. As Hitchcock (2007) has already pointed out, standard neuron diagrams, as in figure 1, have a number of limitations. One thing the dispositionalist will not like is that they represent causal stimulatory connections between discrete events. The causal relations into which these events fall are in no way essential to them whereas a dispositionalist takes the causal relata to be in part constituted by what they are disposed to do (Mumford 2004, ch. 10). But there are shortcomings of neuron diagrams that are less contentious. The frequency and intensity of a cause is not represented. But a book page may break free when it has been turned a thousand times and its first turn contributed just as much to its breaking as the last. Neuron diagrams can also represent only binary relations: a neuron either fires or it doesn't, which is supposed to correspond to an event occurring or not. But an effect can occur with a greater or lesser intensity, depending on the intensity of the cause. Something can be heated to a greater or lesser degree for instance.

But there are two major shortcomings that make neuron diagrams entirely unsuitable for dispositionalists. The first is that it builds in a form of necessitarianism that dispositionalism rejects. If a stimulatory neuron occurs, then

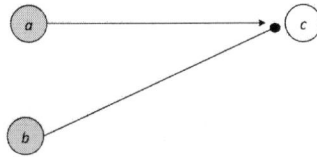

Figure 2: Neuron diagram showing inhibitory connection

its effect has to occur, and we have already said that we wish to replace such a connection with one that is 'only' dispositional. The one exception to this is where an inhibitory neuron fires (as represented in figure 2). In such a case, the effect in question is always inhibited, even if its stimulatory neuron also fires. The inhibitory neuron is thus the only thing that can stop a stimulation, and it necessitates that the effect does not occur. Nature does not work quite that way.

The second major inadequacy for dispositionalists is that an effect only has one cause, whereas we know that most, if not all naturally occurring, causes are complex. Typically, none of these causes would be sufficient on their own for the effect in question. Rather, each will make a relatively small contribution and produce together what none of them could have produced alone. Neuron diagrams cannot represent this, however. A stimulatory neuron, when it fires, is entirely sufficient for the effect, which looks to us to be a major misrepresentation of causation.

The matter is slightly more complicated than that, however, when we consider it in the light of the interpretation of neuron diagrams for counterfactual dependence theories. Lewis, after all, admits that he is looking for a cause, among the others that make the 'total situation' that is followed by the effect. The reason there can be only one stimulatory neuron represented for counterfactual dependence theories is, given that the effect has to occur if the stimulatory neuron fires, that if there were two or more stimulatory neurons, the effect would counterfactually depend on neither. In short, the situation represented would be one in which the effect was causally overdetermined, which has to be denied in the counterfactual dependence theory. The reason Lewis wants only one stimulation for each effect then is not that he denies the complexity of causation, because he doesn't. Rather, he is keen to depict counterfactual dependence as grounding the causal relation. He wishes to depict a situation in which, had the cause not occurred, the effect would not have either. Such a situation exists, however, only if causes cannot be overdetermined. We will state at the end that we think there is no good independent reason to deny overdetermination and thus this limitation on neuron diagrams is not justified. Only if one is already a counterfactual dependence theorist will one accept the limitation of denying ovedetermination and thus neuron diagrams are not after all an ontologically neutral way of representing causation. Rather, a disposition-alist has good reason to look for a better model.

4   Modelling causes as vectors

We have an alternative model to propose: one that better represents causa-tion for dispositionalists and, we will claim, solves some of the problems of Lewis's counterfactual dependence view. This is that causation be modelled as the composition of powers in a vector-like way. The analogy between pow-ers and vectors consists in them both having a direction: powers are all 'for' some type of manifestation. It also consists in them both having an intensity or magnitude as one token power can dispose towards a manifestation more than another one. Intensity is represented by the length of the vector.

We propose to plot these vectors on a one-dimensional quality space.[2] The quality space allows us to show possible changes that powers make possible and frequently produce. Various powers could dispose towards heating or cooling the same room, for instance. From the current temperature, indicated by the central vertical line in figure 3, some powers (a, b and c) dispose to-wards F, making the room warmer, while others (d, e and f) dispose towards G, making the room colder.

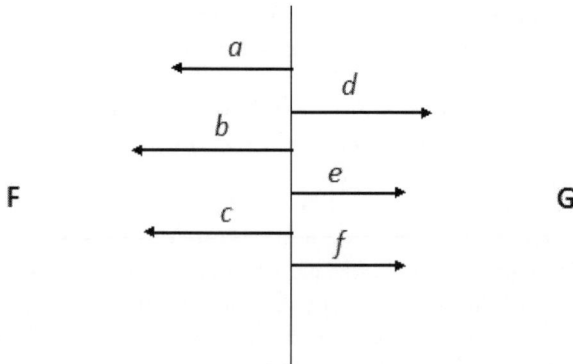

Figure 3: Powers modelled as vectors

There is a further analogy that can be used, which is that of vector addition. The reason that this can be used is that powers can compose, as Mill (1843: III, vi, 3) noted with his idea of the composition of causes. The various individual dispositions at work in some situation can all work together, some towards F

[2]The notion of a quality space comes from Lombard (1986).

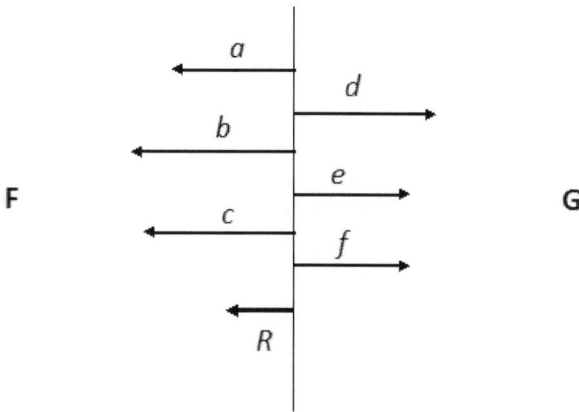

Figure 4: A directed resultant vector R

and some towards G, to make an overall disposition of the situation. In figure 4 we find a case where this happens. The powers disposing towards F are greater than the powers disposing towards G such that overall the situation disposes towards F, indicated by the resultant vector R.

The analogy with vector addition only goes so far, however. There will be some cases where powers compose in the simple additive and subtractive way. In other cases, however, the mode of composition may be more complex. There are some well-known non-linear causal phenomena. One chocolate bar may dispose towards pleasure when eaten but ten, eaten together, dispose towards nausea. And in pharmaceuticals, it is known that two drugs that individually dispose towards health can be harmful when taken in combination. While some powers may add and subtract, therefore, such as forces, and heaters and coolers, the composition may be by a more complex function. The law of gravitational attraction illustrates such a function where the masses and distances apart of two objects produce a force according to an inverse square function, rather than mere addition.

Having noted that complication, however, we can see that the vector model can explain a number of features of causation. In figure 4, the situation overall disposes towards F, as shown by resultant vector R. This itself has to be understood in dispositional terms. It only disposes towards F. F may well be produced, if this disposition is indeed manifested, and when it is produced, it is so by the powers acting together. The vector model thus illustrates polygeny: the idea that an effect is almost always produced by many powers acting together (Molnar 2003: 194–8). *The* cause of F occurring is then understood as the totality of all the powers that were disposing towards F and manifested their disposition in something becoming F. Those powers disposing away from F could still have had some responsibility for *when* something became F or the *degree* of F that the thing became. In the case of the room temperature increasing, for instance, powers disposing towards cooling still had some relevance to the degree of increase. When we consider what it is to be *a* cause of F occurring, we will just focus on the powers that were actually disposing towards F. It would be perverse to call a cause of F something that was actually disposing against it and had to be overcome. F's occurrence was caused in spite of such things, which may thus have limited the degree to which F occurred.

The vector model also allows us to count as cases of causation effects in which nothing happens. There are a number of cases where the effect is not a change or occurrence but a non-occurrence in which, macroscopically at least, nothing changes. Two books leaning at an angle and propping each other up is one instance, another is a fridge magnet sitting motionless on a fridge, and a third is the steady orbit a planet retains around a sun. Dowe (2001) would call these cases of *causation of absence*, given that the effect is a non-change. In the model, these are explained as equilibrium cases, where powers are at work but they balance out (figure 5). For the realist about powers, there is a huge difference between the case in figure 5 and a case where there are no powers at work, even though the result may be the same: nothing happens. Hence, in a tug-of-war, the two sides may be equally balanced and the rope thus progresses neither east nor west. Powers are at work, however, and

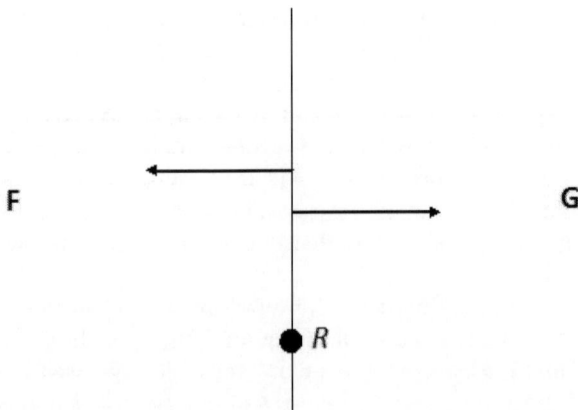

Figure 5: A zero resultant vector where powers are in equilibrium (causation of absence)

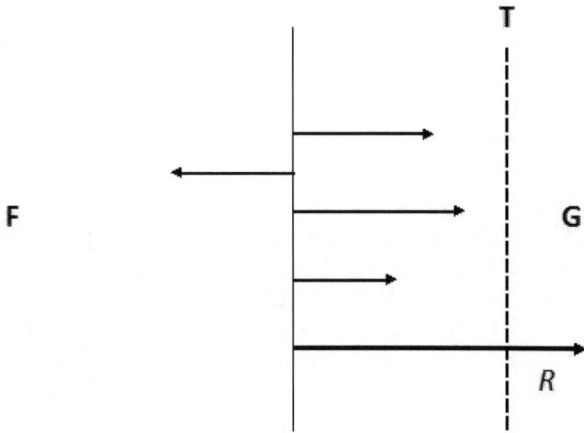

Figure 6: A resultant vector that passes threshold T

thus the case differs from another one where the rope does not move: namely where no one is pulling it.

The employment of a notion of threshold is also of use. There are some cases of causation where at a certain point some novel or particularly interesting phenomenon occurs. Water can be heated to various temperatures, for instance, but at 100°C we get the significant phenomenon of it turning to steam. This is a point of discontinuity where a significant new process is triggered. We may therefore want to represent such a threshold, T, on our vector diagram (figure 6). Such thresholds are often important to us as causal agents. Our desire is often that some effect be brought about and in our deliberative actions we are trying to assemble enough for it. In striking a match, for instance, I am seeking that it lights. I know that I have to produce adequate friction against the match box, with a match that is dry enough, which has its flammable tip intact and is kept out of the wind.

We can broaden out from the last point and note that vector diagrams could be used in decision theory. Simply plot out all the relevant considerations according to the directions in which they point and their importance, indicated by their length. We do not develop that idea further here but it does bring us to an important point. When we make such decisions, we sometimes have just two choices and we will often have an overriding desire that one of those two choices be made. I may be invited to both Sydney and Auckland on the same day, for instance, and if I am equally disposed towards each it might indicate that I end up in the sea.[3] But in such a case, even if I find it hard to

---

[3]This objection due to Mark Colyvan, in discussion.

split the two, I would rather go to one of them rather than none. This means that I will allow only one of two outcomes, even if there is only slightly more in favour of one than the other. And if there is nothing at all that disposes overall towards one outcome, I would rather toss a coin than make no decision. This decision-theoretic case can, however, be reflected in natural processes. The vector diagrams presented thus far were designed to allow effects that would admit of degree. But not all do. F and G may not concern magnitudes but may be simple states: effectively like off and on switches. Whether there is overall a lot or just a little that disposes towards the switch being on, it is simply on nevertheless. Effectively, the threshold for being on and off is the same line and what counts is which side of that line we end up. The neuron model was criticised for depicting all effects as simple binary, all or nothing cases. In allowing degrees of an effect, we do not want to make the mistake of denying that any effects have this binary nature.

5   Cartwright versus Mill on component powers

There is a threat to the account presented so far that comes from a surprising source. Although Cartwright is a supporter of causal powers or capacities, she also articulates the empiricist argument against the reality of component forces. There are good reasons to believe in resultant forces, because they are measurable, but an empiricist can only be instrumentalist about the components (Cartwright 1983: ch. 2). Cartwright's discussion concerns forces and it should be clear that causal dispositionalism is not an attempt to reduce all powers to physical forces. A decision between reductionism and holism is another matter, on which dispositionalism remains neutral. What is important here, however, is that Cartwright's argument could be applied to powers as well as forces. Might it be said that only the resultant powers are real and that we should be instrumentalist about the components?

Cartwright's concern is with whether vector addition is an accurate story for the composition of forces and, for our purposes, causation in general. She is sceptical, thinking that we do the addition, not nature. She explicitly rejects Mill's view on composition of causes in which all the component causes exist in the overall composed cause:

> In this important class of cases of causation, one cause never, properly speaking, defeats or frustrates another; both have their full effect. If a body is propelled in two directions by two forces, one tending to drive it to the north and the other to the east, it is caused to move in a given time exactly as far in both directions as the two forces would separately have carried it ... (Mill 1843: III, vi, 1, p. 370–1).

Cartwright (1983: 60) quotes this in order to deny it. The body makes no movement north at all, nor any movement east, just a single movement north-

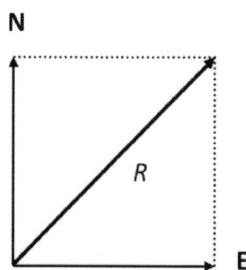

Figure 7: Mill's example of a resultant force

east (figure 7). And if we had bodies forced at an obtuse angle to each other, the resultant would not move as far in one direction (west) as it would have done had there not also been the force north by north east (figure 8). Worse still, Cartwright cites the example of counterbalancing north and south forces acting on a body, which does not then move at all (1983: 61). How can Mill say that each force has its full effect if a northerly force and southerly force produce no movement at all?

There is an answer from Mill himself, however, which can be found just if we continue the sentence where we left off in the previous quotation: 'it is caused to move in a given time exactly as far in both directions as the two forces would separately have carried it; *and is left precisely where it would have arrived if it had been acted upon first by one of the two forces, and afterwards by the other.*' (Mill 1843: III, vi, 1, p. 370-1, emphasis added) Thus, it is as if one force first did its work and then the other did. Even in Cartwright's case

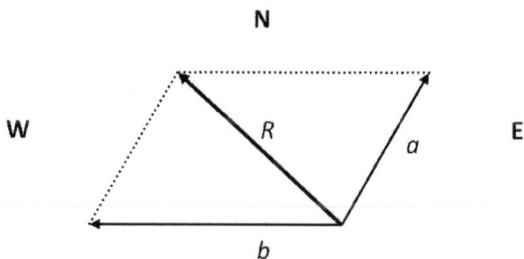

Figure 8: An example not considered by Mill

of a body subject to opposing balanced forces, resulting in no movement, the body is left where it would have been had first one, and then the other, moved it. And in one very important sense, both forces have their effect. For if we consider counterfactually, what would have happened had one of them been absent, then we can see that their presence made a difference.

This is what we want to say about all the component causes in a combined cause. All of them make a contribution to that combined cause. They tend to make a difference, setting aside cases of overdetermination, which will be considered in the next section. There are cases where they come together and get subsumed in the resultant power of the situation, but they nevertheless retain their reality. It is the individual powers that drive causation, not the resultant ones. Both the components and resultants can be understood as causally active, however. The resultant and its components are not wholly distinct existences so we can allow that they do the same work without it being a case of overdetermination. But it is the components that make up the resultant, rather than the other way round. Given the components, there can be only one resultant, but we could not derive from a resultant what its components were. It is best to take these components as real in themselves.

6   A return to counterfactuals ... with an unexpected pay-off

We return now to the issue of counterfactuals. We have introduced an alternative way to neuron diagrams for modelling causal situations. But Lewis's possible worlds also ground an account of counterfactuals and counterfactual dependence is a key notion in his theory of causation. Can we account for causal counterfactuals?

There is a very simple answer to this, though unfortunately we cannot quite agree with it. The simple response would be to say that real, worldly causal powers were the truthmakers of the counterfactuals. There is, after all, a strong counterfactual intuition when we have causation. We think causes should make a difference such that had they not occurred, something would have been different: the effect would not have occurred, for instance. A power could be the truthmaker of a counterfactual because it is precisely the difference maker. In a standard case, it contributes to an effect such that, without it, the effect would not have occurred. What is in mind is depicted in figure 9. This vector model depicts a case in which there is a resultant vector directed towards F, and let us assume that this resultant power manifests in something becoming F. However, we can see that had component power b not been present, which is indicated by the vector being a dotted line, overall the situation would not have disposed towards F but would have been in equilibrium. The suggestion would be, therefore, that we are entirely fictionalist about the counterfactual situations themselves. They have no existence,

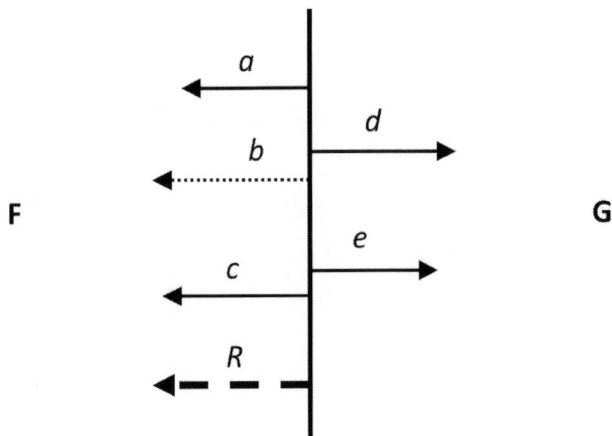

Figure 9: Vectorial modelling of a counterfactual

whereas they do in Lewis's metaphysics, albeit in other worlds. The counterfactual situations we model are entirely in our imagination. But there are nevertheless worldly truthmakers of the counterfactual truths in the real powers
that cause effects.

The unexpected pay-off of this account of counterfactuals is that it can also
be used as our account of causation by absence, although with one important
difference. Causation by absence, if it is real, could be a problem for our account. We say that powers do all the work. But a genuine absence is nothing
at all (it is not a *something* by another name, for instance). And an absence
cannot bear any powers, for that would be like them floating freely and not being powers of something. An account is not hard to find, however. In cases
of causation by absence, we maintain, a component power is removed not
just in the imagination but in reality. In a case where we forget to water the
plant, we might say that absence of water caused its death. But this cannot be
quite right. Absent water is nothing and cannot cause anything: that would
be *creation ex nihilo*. What has happened is that the plant was in equilibrium,
avoiding the twin perils of death by drowning and death by dehydration. The
surrounding atmosphere has a power to suck moisture out of the plant's leaves
and soil but this is counterbalanced by the addition of new water. When I cease
to water the plant, it is not the absent water that kills it: it is the dehydrating
power of the atmosphere. This is depicted in figure 10. Power *b* – the hydrat-
ing power of water – is removed and its removal is the occasion for the plants
death. But it does not cause its death. The causes are the powers *c* and *d* that

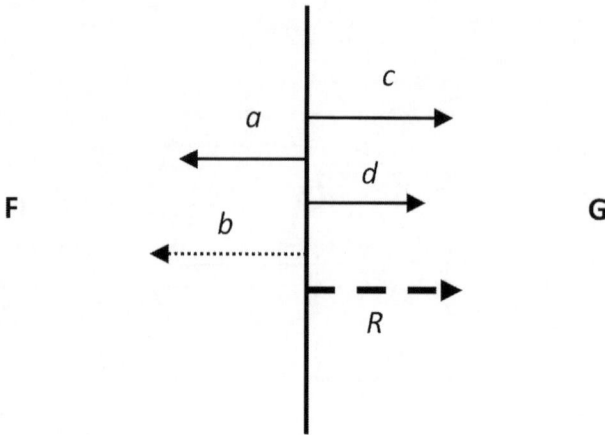

Figure 10: 'Causation by absence'

dispose towards its dehydration. Once *b* is removed, the situation moves from an equilibrium to a directed resultant vector, R.

That would be a good place to finish but we cannot do so without offering two major amendments to our theory of counterfactuals. One concerns the dispositional modality and the other concerns overdetermination. The point about dispositional modality concerns counterfactuals in which some power is added to the actual situation; while the point about overdetermination concerns counterfactuals in which a power is subtracted from the actual situation.

The counterfactual dependence view of causation is that A causes B where A and B both occur and had A not occurred, B would not have occurred either. The situation we are fictionalising is one without A but, of course, many causal counterfactual claims will be about what would have happened had some additional power been present: for instance, if there had been a spark, we would have had an explosion. If we accept that causation employs the dispositional modality, we will have to recast such counterfactuals. We cannot say that the explosion would definitely have happened in such a situation. Rather, we will have to interpret such a counterfactual as meaning that had there been a spark, there would have been a tendency towards an explosion. The spark would have disposed towards the explosion, probably made it more likely, and so on. Given causal dispositionalism, necessity is jettisoned. A causal counterfactual that concerns additional powers therefore has to be recast slightly. But is this really to be considered a price of causal dispositionalism? Arguably not because the recast counterfactual more accurately reflects the fact that powers only tend towards their manifestations, without guaranteeing them, and this is

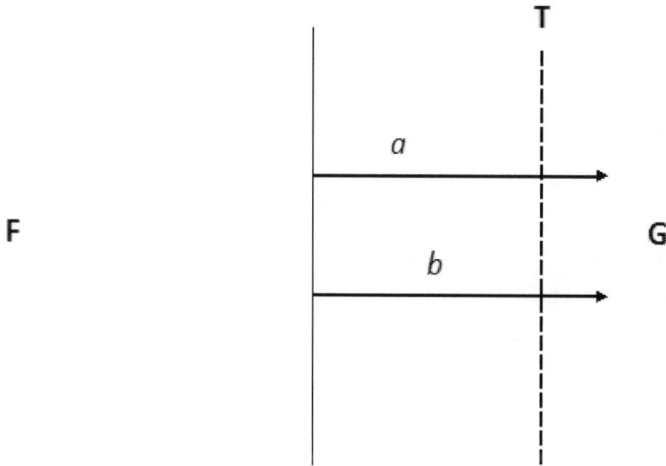

Figure 11: Causal overdetermination

how we think the world works. A Lewisian might object that if we hold all else stable, and just add one further power, then the effect has to follow. But we reject this assumption (Mumford and Anjum 2010). A disposition could always fail to manifest, even if for no good reason.

The counterfactuals from which something is subtracted, where A does not occur for instance, face the notorious problem of overdetermination. If an effect e can be overdetermined by causes a and b, each of which is enough on its own for e, then e counterfactually depends on neither. Counterfactual dependence theorists of causation therefore have to deny overdetermination. Otherwise, there could be cases in which a cause was removed and the effect would still occur just the same. Having no such vested interest in the matter, we can grant the possibility of overdetermination. A room could contain two thermostatically controlled heaters, each programmed to gradually raise the room temperature to 21°C within 10 minutes of being switched on. Someone switches them both on and they do their work. Had one of them not been turned on, the effect would have been the same. We need to go to no lengths to argue away this kind of case as we can accommodate overdetermination, as illustrated in figure 11.

Where does that leave our counterfactuals? One thing it means is that the counterfactual intuition – which is a difference-making intuition – does not apply in every case of causation. Some causes don't make a difference, where they are overdetermining or redundant causes.[4] Counterfactual dependence

---

[4] For a taste of redundant causes, see Schaffer's (2000) trumping pre-emption cases.

STEPHEN MUMFORD AND RANI LILL ANJUM

can then be used as a diagnostic of causation, allowing us to identify most cases, but it is not an infallible indicator of causation and certainly not in a position to be constitutive of causation.[5] This last conclusion of course is one that should lead us to look for alternative theories of causation and enough evidence has been presented here, we hope, to make causal dispositionalism a realistic contender.

---

[5] For a case of counterfactual dependence between events but without causation, see Mumford and Anjum 2009.

POWERFUL PROPERTIES AND THE CAUSAL BASIS OF DISPOSITIONS

Max Kistler

1   Introduction

Many predicates are dispositional. Some show this by a suffix like '-ible', -uble', or '-able': sugar is *soluble* in water, gasoline is *flammable*. Others have no such suffix and don't wear their dispositionality on their sleeves. Yet part of what it is to be solid is to be disposed to resist deformation, and part of what it is to be red is to appear red to normal human observers in normal lighting conditions.

However, there is no agreement as to whether dispositional predicates may be given a realist interpretation. For many authors, propositions containing them are made true by states of affairs (or facts) containing categorical, rather than dispositional properties. Many also claim that the states of affairs that make true attributions of dispositions to macroscopic objects are microscopic states of affairs concerning their parts. For example, what makes a vase fragile is the microscopic structure of its molecular constituents, which is what makes the vase break when it falls.

Against these claims, I will argue that what makes a dispositional predicate apply to an object, whether macroscopic or microscopic, whether in common sense or science, is the object's having what I will call a powerful property. If the object is macroscopic, it is another matter whether the property is microreducible. My reasons for supposing that these powerful properties exist are those for postulating theoretical properties generally: they unify existing explanations and suggest new ones.

My plan is as follows. I will begin with a brief sketch of the debate between what I call the reductionist and the realist doctrines about dispositions.

Then I will argue that there are many so-called multi-track dispositions, both in common sense and in science, and that accounting for them requires a distinction between the concepts of powerful property and disposition. I will give three examples of multi-track dispositions, one from common sense, one from physics and one from cognitive psychology. In each case in which an object has a multi-track disposition, it has a powerful property that contributes causally to bringing about the manifestations of the disposition. The distinction between powerful properties and dispositions makes realism and reductionism compatible: Realism is justified with respect to powerful properties, whereas the ascription of a disposition is reducible: it is made true by the fact that the object has a powerful property, together with laws of nature. I will then justify the existence of real powerful properties against several arguments: that such properties are pseudo-properties having only verbal existence, that only their reduction base is causally efficacious and therefore real, that some dispositions need no such causal basis, and that there are no irreducible multi-track dispositions in the first place.

2   Reductionism and Realism

Take a truthful attribution of the dispositional predicate 'is elastic' to a rubber ball.

(1) This rubber ball is elastic.

Reductionists hold that such attributions of dispositions can be analysed in non-dispositional terms. Carnap (1936, 1937) has shown that it is impossible to analyse them within standard first order logic, using in particular the material conditional. Goodman's (1955) thesis that they can be analysed in terms of counterfactual conditionals instead, has launched a rich debate, which I will not go into here.[1]

Rather, I will consider the thesis, put forward by Quine (1971) and Armstrong (1973), that attributions of dispositions can be reduced in another sense, i.e. in the sense of reduction between theories. According to Quine (1971: 10), the role and utility of attributions of dispositions is to 'refer to a hypothetical state or mechanism that we do not yet understand'. In Armstrong's words, 'dispositional concepts leave us in ignorance concerning the properties of the disposed object which give it that disposition' (Armstrong 1973: 417). If the attribution of a dispositional predicate to an object is true, there must be some state of affairs involving the object that makes the attribution true.[2] Armstrong suggests that these truthmakers are microscopic: In the case of brittleness, the underlying microscopic state is 'a certain sort of bonding of the molecules of the brittle object' (Armstrong 1973: 417). Scientific research

---

[1] See Mumford (1998), Gnassounou and Kistler (2007).

[2] The theory of truthmaking is developed in Armstrong (2004).

leads to a 'contingent identification' (Armstrong 1973: 420) of the disposition with the microstate. The difference between the disposition and its microscopic reduction basis is epistemic. Ontologically speaking, the disposition just *is* the reduction base, vaguely or incompletely conceived.

Prior, Pargetter, and Jackson (1982), who share the reductionist approach to dispositions, deny that dispositions are identical to their reduction basis. They agree with Quine and Armstrong that if an object can truly be said to have disposition D, then it must have a first-order state L that is causally responsible for the manifestations E, which are characteristic of D in appropriate triggering circumstances T. But they argue that the disposition D cannot be identical with this first-order state L, which they call D's 'causal basis' because D has, or at least can have, different causal bases in different objects.

In spite of this difference, all reductionist accounts we have mentioned share the idea that every true attribution of a disposition has a non-dispositional truthmaker: a state of affairs, generally taken to be microscopic, or microstructural, involving a categorical property.

I will call 'realist' those accounts of dispositions, which deny that attributions of dispositions must have categorical truthmakers.[3] Realists argue that at least some attributions of dispositions have truthmakers that contain a dispositional, rather than a categorical property. For realists, dispositions are properties in their own right, which can exist without a so-called 'categorical basis'. Realists typically defend the following theses:

1. Attributions of dispositions cannot be analysed in non-dispositional terms. What makes true the proposition that this ball is elastic is the fact that the ball has a real powerful property of elasticity.

2. Even if some dispositions actually have a basis, they need not have one. It is an empirical issue whether macroscopic powerful properties such as elasticity can be reductively explained in terms of the properties of the microscopic parts of elastic objects. The existence of the macroscopic disposition does not depend on the possibility of such a reduction.

3. Some powerful properties have a reductive basis that is itself dispositional.

4. Attributions of dispositions to elementary particles cannot be reductively explained. If there are true attributions of dispositions to elementary particles, their truthmaker must contain a fundamental powerful property that has no basis.

One important thesis of this paper is that the reductionist and realist positions are only incompatible insofar as one conflates two concepts: *powerful*

---

[3] Realist accounts have been put forward by Harré and Madden (1975), Mumford (1998), Ellis (2001), Molnar (2003), Bird (2007a). Mellor's (2000) account does not fit in this classification. He takes temperature to be a 'real disposition' that causes the manifestations of temperature, but not fragility, because different properties cause the manifestations of fragility in different things.

properties (PP) and *dispositions.* My argument will be based on the analysis of so-called 'multi-track' dispositions. The distinction between powerful properties and dispositions is required to account for multi-track dispositions such as elasticity. In light of this distinction, it appears that realism and reductionism can be reconciled: Realism is right about powerful properties, whereas reductionism is right about dispositions.[4] The distinction between powerful properties and dispositions requires the notion of a law of nature. Each law in which a given powerful property plays a role corresponds to a disposition of the objects possessing the powerful property.[5]

3   Multi-track dispositions

Gilbert Ryle has observed that many dispositional predicates are tied to a whole set of triggering-manifestation pairs. Such dispositions are often called 'multi-track' dispositions. Being elastic is a traditional example of a common sense dispositional predicate whose meaning is related to several manifestations. 'We can say that something is elastic, but when required to say in what actual events this potentiality is realized, we have to [...] say that the object is contracting after being stretched, is just going to expand after being compressed, or recently bounced on sudden impact.' (Ryle 1949: 113). Common sense treats elasticity as one property, not several. However, to be elastic entails having several dispositions, in the sense that several conditionals are true of every elastic object; and these conditionals are typically irreducible to one another. If an object o is elastic, then 1) if *o* is stretched, then *o* will retract once released; and 2) if *o* is compressed, *o* will expand once released, etc. As Mumford remarks, 'an ability to bounce (when dropped) is different from an ability to bend (when pressured) though both might reasonably be thought dispositions of something that is elastic, in virtue of its elasticity' (Mumford 2004: 172).

Scientific dispositional predicates too are semantically linked to several triggering-manifestation conditionals. Consider electrical conductivity σ. The concept of electrical conductivity is introduced to account for the fact that different materials react with electrical current of different intensity to a given electric field. The conductivity of a given piece of copper can manifest itself in infinitely many ways: by giving rise to electric current of density J_1 in a situation with electric field E_1, to J_2 in the context of electric field E_2, etc. All these dispositions to manifest by current densities J_i, given field E_i, can be expressed by a single law stating that J is proportional to E, $J=\sigma E$. However, electrical conductivity is not just shorthand for the ratio J/E. Rather, it is a theoretical

---

[4]As Shoemaker (1980) and Mellor (2000) have shown, the dispositional-categorical distinction applies strictly speaking to predicates not properties. Cf. Kistler (2007). See below, Section 6.3.

[5]It is noteworthy that, if it is correct that powerful properties must be distinguished from dispositions, we have a reason to believe that there are laws, against Mumford's (2004) claim that dispositions make laws metaphysically superfluous.

property. It figures in laws relating it to other properties. As with other theoretical properties, postulating its reality allows explaining the links between σ and various other properties. According to one of these laws, the Wiedemann–Franz law, the thermal conductivity λ of metals is proportional to the electrical conductivity σ.

$$\lambda = \sigma \frac{1}{3} \frac{\pi k_B}{q} T$$

where T: temperature, k_B: Boltzmann's constant, q: unit charge.

All metals possessing electrical conductivity σ have thermal conductivity λ. Thermal conductivity is itself a powerful property that gives conductive substances such as copper different dispositions to manifest in different triggering circumstances. Given that σ is lawfully linked to λ, the fact that an object has σ gives it indirectly all the dispositions of λ. σ manifests itself also in the manifestations of λ; in particular, σ can be measured by the manifestations of λ. In this way, the manifestations of σ are multiplied by the laws which link σ to other properties F, and then again by the laws which link these F to still other properties, and so on. Each of these manifestations corresponds to a different disposition that can be truthfully ascribed to an object by virtue of its possessing σ.

A person's iconic memory of a stimulus is a cognitive multi-track disposition. Iconic memory is a form of short-term memory—it decays rapidly, lasting for approximately one second—that has been postulated by Sperling (1960) to account for a surprising phenomenon connected to a set of behavioral dispositions.

7	1	V	F
X	L	5	3
B	4	W	7

Figure 1: Stimulus of one of Sperling's experiments (from Sperling 1960: 3)

In Sperling's 'partial report condition', subjects are trained to report selectively the content of the upper, lower or middle row of the array, according to whether they hear, immediately after the presentation of the stimulus, a high-middle- or low-pitched tone. Immediately after the end of the exposure to the stimulus, subjects are capable of reporting on average three items from each of the three rows, although they cannot report them all in the same trial. This capacity can be explained by postulating the existence of iconic memory, in which subjects retain, for about one second, about nine characters from a twelve character array.

Having a given content in one's iconic memory is a powerful property. It is not directly observable; however, the existence of a memory state containing nine items explains three dispositions the subject has immediately after exposure to the stimulus, such as: If she hears a high frequency tone, she reports on average three items of the top row.

Both common sense and scientific discourse provide cases where having a powerful property (being elastic, having an electrical conductivity of 60×10^6 Siemens/meter, having stored nine letters in iconic memory) entails having several dispositions. However, as long as 'powerful property' and 'disposition' are not distinguished, we seem to be caught in the paradox of having to say of the same property instance that it is both one and several. Consider an elastic rubber ball b and suppose we have only one concept of disposition. The following three statements are true of b.

(1) b has the disposition of elasticity.

(2) b has the disposition to expand after having been compressed.

(3) b has the disposition to retract after having been stretched.

(1), (2), and (3) all seem to ascribe the same disposition: elasticity, as opposed to, say, opacity or inflammability. However, (2) and (3) cannot attribute the same disposition. (2) attributes to b the disposition to expand (after having been compressed) and (3) the disposition to retract (after having been stretched). A disposition to get longer cannot be identical with a disposition to get shorter.

Distinguishing between powerful properties and dispositions yields a straightforward way to avoid the paradox. A disposition is defined by a specific conditional linking a triggering condition to a manifestation. The disposition D_1 to retract (manifestation M_1) when released after having being stretched (triggering condition T_1) is different from the disposition D_2 to expand (manifestation M_2) when released after having been compressed (triggering condition T_2), and also different from the disposition D_3 to bounce (M_3) when dropped (T_3). Elasticity is the powerful property that gives object b its dispositions D_1, D_2, D_3. Now, what exactly does 'give' mean in this context? What is the relation between the property P of being elastic and the different dispositions D_i it gives its bearers?

The simplest hypothesis is that P is identical to D_i, for all i. But this cannot be correct, because the dispositions D_i are not identical with each other. Another hypothesis is that P is identical to one specific disposition D_j, but not to the others. This does not fit our case, because elasticity is not linked more closely to one of these dispositions than to the others.

Ryle's thesis is that the relation between a predicate expressing a multi-track disposition such as elasticity and a predicate expressing a single disposition is equivalent to the difference between a *determinable* predicate and one of its *determinates*. 'Some dispositional words are highly generic or determinable, while others are highly specific or determinate.' (Ryle 1949: 114) The predicate 'x is a baker' is determinate in the sense that it names a disposition with a unique manifestation, the activity of baking. 'x is a grocer' is less determinate, in other words, a determinable. The dispositions to act of someone satisfying this predicate are expressed by the determinates of this determinable: 'x is selling sugar now', 'x is weighing tea now', 'x is wrapping butter now' (Ryle 1949: 114).

However, the distinction between determinable and determinate is not the same as the distinction between a powerful property and the dispositions it gives its bearers. All properties that are determinates of a given determinable are of the same type: Either they are all intrinsic, or they are all relational. However, a powerful property and the dispositions it gives its bearer do not belong to the same category in this respect. A powerful property is an intrinsic property, whereas a disposition is relational: The identity conditions of a disposition depend in an essential way on circumstances which are in part external to the object that has the power, and in particular on the laws of nature. In general, neither the triggering circumstances T nor the manifestation M are intrinsic properties of the powerful object. That a determinable and its determinates are properties of the same type is a consequence of a general constraint on the relation between them. As Funkhouser (2006) explains, the determinates of a given determinable determine it according to a limited number of 'determination dimensions'. For colours, there are, e.g., three such dimensions, hue, brightness and saturation, so that a given colour C corresponds to a subspace S of the three-dimensional colour space. Determinates of C correspond in turn to subspaces of S. A relational property such as a disposition cannot correspond to a subspace of the space corresponding to an intrinsic property, such as a powerful property.

Here is my hypothesis about the relation between a powerful property P and the dispositions D_i it gives its bearers. Ascribing disposition D_i to object b is equivalent to asserting the counterfactual conditional 'if b were in conditions T_i, then ceteris paribus, b would manifest M_i'.[6] It is justified to ascribe D_i to the object b, rather than to the circumstances T_i, if and only if there are intrinsic properties of b by virtue of which b contributes causally to the production of M_i. Let us call the 'causal basis' in b, of the disposition D_i to manifest M_i, the set B_i of those intrinsic properties of b that intervene causally in the production of M_i, alongside the triggering circumstances T_i and other background conditions. In general, each D_i will have its own causal basis B_i. A given set of dispositions D_i is grounded on a unique common powerful property if and only if the intersection $I = \cap_i B_i$ is not empty. In that case, the powerful property underlying a given set of dispositions D_i is the conjunction of the properties belonging to I. In sum, the powerful property underlying a set

---

[6]The ceteris paribus clause is required to account for exceptional circumstances, such as the presence of masks or antidotes. Cf. Johnston (1992), Molnar (1999), Bird (1998).

of dispositions D_i is the conjunction of those intrinsic properties of the object to which the dispositions D_i are truthfully ascribed, which contribute causally to bringing about all of their manifestations M_i. There is, e.g., a real property of elasticity belonging to the rubber ball b if and only if there is a set of intrinsic properties of b, which contribute causally to the manifestations of all the dispositions associated with elasticity: bouncing after being dropped, retracting after having been stretched, etc. Conceived in this way, powerful properties have the status of theoretical properties, in the sense that the postulate of their existence is fallible, and has to be justified by the usual criteria of theory construction: It must enhance the simplicity of explanations and predictions in general and causal explanations in particular and it must be fruitful, in the sense of suggesting new explanations and predictions.

If single-track dispositions exist, they are a special case of multi-track dispositions, in which the causal basis of a given disposition has no elements in common with the causal bases of any other dispositions.

4   Arguments for powerful properties

The hypothesis that there is a real powerful property underlying a set of dispositions D_i, has the form of an inference to the best explanation. If there is a unique powerful property that is the causal basis of a set of dispositions D_i, it provides a unique and unifying explanation of their manifestations M_i.

In electric fields E_i of various strengths a copper wire w manifests different dispositions to produce different current densities J_i. Electrical conductivity σ is postulated as an intrinsic property of w that contributes causally to bringing about the different J_i in different fields E_i. The existence of this common causal basis makes it possible to express the relation between the different E_i to the J_i in a single equation.

As a theoretical property, a powerful property may lead to explanatory unification at a larger scale. σ figures in various theoretical laws, such as the Wiedemann–Franz law. These laws express relations between σ and other theoretical properties, which figure themselves in still other laws. Thus σ mediates, and helps to explain, relations between other theoretical properties to which σ is linked, directly or indirectly, by laws. Electrical conductivity being linked both to thermal conductivity (by the Wiedemann–Franz law) and the magnetic field B (via the law linking J_i to E_i, and the laws linking E to B, such as Ampère's law and the Maxwell–Faraday equation), it can explain systematic correlations between λ and B, which might have appeared mysterious without σ. In a similar way, the content of the iconic memory of Sperling's experimental subject is a theoretical property that explains in a simple and unified way, several of her dispositions to behave in the different partial report conditions.

A second reason for postulating powerful properties is, as for other theoretical properties, their fruitfulness. The postulate is justified if it suggests new experimental hypotheses. This seems to be the case for iconic memory. It has been introduced to account for the performance of subjects in the partial report conditions for different lines of the array. However, iconic memory suggests new testable hypotheses about performance in yet unexplored partial report conditions: Subjects should have dispositions to report columns and diagonals of the array.

5 Arguments against efficacious powerful properties

5.1 Powerful properties are pseudo-properties

Reductionists will object that the concept of a property that is both dispositional and a causal basis, and thus causally efficacious, is contradictory. According to one traditional argument, dispositions are not real because they are not causally efficacious. Dispositions, so the argument goes, are like the famous 'dormitive virtue' mentioned in Molière's play 'The Imaginary Invalid' (also called 'The Hypochondriac'). When asked for the cause of opium smokers' sleepiness, Molière's candidate (for becoming a medical doctor), mentions the dormitive virtue of opium. This is supposed to ridicule pseudo-explanations in scholastic style. Indeed, the bachelor's answer seems to be a tautology rather than an explanation. If we mentioned dormitive virtue to explain sleep we would 'expose ourselves to Moliere's ridicule, and, if we did nothing further, we would deserve it.' (Armstrong 1973: 419) The reason is that the expression 'dormitive virtue' is analytically linked to sleep, meaning nothing more than 'property that causes sleep'. Now it can be argued that this shows that dormitive virtue cannot cause sleep, for causation is contingent.

However, such an argument is fallacious because it fails to distinguish between a property and the descriptions of a property. The 'dormitive virtue' argument against the efficacy of powerful properties can be criticized in a way analogous to Davidson's rebuttal of an argument against the thesis that our actions are caused by our reasons. Davidson (1980: 14) points out that if c is cause of e, then we can truthfully name c 'the cause of e'. The statement 'the cause of e caused e' is certainly tautological, but it would be mistaken to take this to show that the cause of e, i.e. c, didn't cause e. 'The cause of e causes e' (for an event e) is true even though cause and effect are described in such a way that the resulting statement is analytic.

Here is the analogous reasoning concerning the efficacy of properties. Assume it is true that 'the property of bearing electric charge q creates, at distance d, an electric field of strength E'. Replacing the expression 'property of bearing electric charge q' by the description 'property that creates, at distance d, an electric field of strength E' makes the statement analytic: 'the property that creates, at distance d, an electric field of strength E, creates, at distance d, an electric field of strength E'. However, this does not deprive the property of bearing electric charge q of its causal efficacy. The same fallacy seems to be operative in the case of the dormitive virtue of opium. The property of opium that

is efficacious in bringing about sleep when absorbed in the way X can truthfully be named 'the property that (ceteris paribus) brings about sleep when
absorbed in the way X'. The fact that we produce an analytical statement by
calling it that way when stating the cause of sleep, does not deprive this property of opium of its efficacy in bringing about sleep.

Opium is not a placebo.[7] The problem stems from the functional meaning
of the expression 'dormitive virtue'. It determines the identity of the property
only indirectly, by its typical effect. This does nothing to put in doubt its reality
or causal efficacy.

5.2   Macroscopic dispositions are not real, only their microbases are

Another influential argument against real efficacious powerful properties has
already been mentioned above. Prior, Pargetter, and Jackson (1982) argue that
1) all dispositions have a causal basis, 2) this basis is distinct from the disposition itself and 3) all causal efficacy erroneously attributed to the disposition
really belongs to the basis.

Contrary to Armstrong and Quine, Prior, Pargetter, and Jackson refrain
from arguing that the causal basis must be categorical.[8] However, they identify the *causal* basis of a macroscopic disposition, i.e. the intrinsic property
by which the bearer of the disposition contributes to bringing about the manifestation, with its microscopic *reduction* basis.[9] Typically, the reduction base
consists of some microscopic property such as 'having molecular bonding *a*',
or 'having crystalline structure *b*' (Prior et al. 1982: 253). From the premises
that (1) the causal basis is the microscopic reduction basis and that (2) it is distinct from the macroscopic disposition, they conclude that the macroscopic
disposition is not real, i.e. not efficacious. Armstrong and Quine come to the
same conclusion: For them, the disposition is identical with the basis, but the
basis is categorical, and a property cannot be both categorical and dispositional. Therefore, there is no real macroscopic disposition.

These arguments against the reality of macroscopic dispositions share the
same weakness: they do not justify the implicit presupposition that the causal
basis is necessarily identical with the microscopic reduction basis. If this presupposition is false, Armstrong's, Quine's and Prior, Pargetter, and Jackson's
arguments do not refute the thesis that macroscopic powerful properties such

---

[7]We shall return to placebos below, in section 6.1.

[8]'We have not here argued that the causal basis is categorical or intrinsic, only that it exists.'
(Prior et al. 1982: 253).

[9]Armstrong and Quine do not claim that the causal basis of a macroscopic disposition is necessarily microscopic, but they suggest that this is at least the typical case. When we attribute a
disposition to *x*, says Quine (1966: 73), 'we are attributing to *x* some theoretical explanatory trait
or cluster of traits. Typically these would be traits of microscopic structure or substance.' For Armstrong (1973: 417), 'attribution of a disposition to an object entails that the object is in a certain
state. What is the concrete nature of this state? [...] In the case of brittleness, for instance, the
state will be a certain sort of bonding of the molecules of the brittle object.'

as electrical conductivity, the iconic memory of a stimulus, and even fragility can be the causal basis of their manifestations.

I cannot pretend to establish here that macroproperties can be causally efficacious, whether or not they are microreducible. But I can make this claim plausible by looking at a simple example of the microreduction of one paradigmatic dispositional macroproperty, i.e. electrical conductivity. The analysis of this case shows that the reasons to think that the microscopic reduction base is more apt as a causal basis are ill founded. Indeed, the reasons for this conviction are the hypothesis that the properties in the reduction base are 1) purely categorical and 2) purely microscopic. However, as we will see, the properties to which electrical conductivity is reduced are neither purely categorical nor purely microscopic.

Electrical conductivity of metals can be reductively explained in a simple model due to Drude. Although it is classical and does not take into account the quantum mechanical structure of metals, the Drude model provides a good approximation in a large range of common circumstances.[10] The model is built on the hypothesis that electrical current is due to the motion of 'free' electrons, i.e. electrons moving freely through the crystal grid constituting the metal, and to the electric charge carried by these free electrons. The macroscopic electrical conductivity of metals is explained in terms of the following properties of the microscopic constituents of metals: the number of free electrons per cubic centimetre n, the charge of an individual charge carrier q, the so-called relaxation time or mean free time of the free electrons, i.e. the average time interval between two collisions, τ, and the mass of an electron m.

The bridge law (1) expresses the core of the reduction.

$$\frac{nq^2\tau}{m} = \sigma \quad (1)$$

(1) shows that the reductive basis of σ lies in microproperties of the metal's free electrons. Now, the microproperties mentioned in the reduction of conductivity are neither purely categorical nor purely microscopic. Here is why they are not purely categorical. The expression on the left hand side of equation (1) is equivalent to the product of the mobility μ of an individual electron with its charge q and the number n of electrons per unit volume.

$$nq\mu = \sigma$$

The dispositional character of the mobility μ of electrons is apparent in its name. It is the disposition to move with speed v in external field E. This shows that the mere fact that a property is microscopic, and used in a reductive explanation of a macroproperty, is not sufficient to establish that this property is categorical. Moreover, as several authors have argued,[11] the properties of

---

[10]Quantum mechanical models of conductivity are required to explain phenomena occurring in extreme conditions, such as supraconductivity.

[11]Popper (1957), Mumford (1998), Ellis (2001), Molnar (2003), Bird (2007a).

having elementary charge q, mass m, and mean free time τ, also have a dispositional essence, although this is not apparent in their name. To bear charge q just is to have a set of dispositions, such as the dispositions to attract charges of opposite sign, to repel charges of equal sign, and to accelerate along external fields. Similarly, to have mean free time τ is to have the disposition to move without collision for a mean time of τ seconds.

Paradoxical as it may sound, some of the properties in the basis of the microreduction of conductivity are also not purely microscopic. Properties such as mobility μ and mean free time τ are microscopic insofar as they are truly attributed to microscopic particles, i.e. electrons; however, they are not purely microscopic in the sense that they are dependent on, and determined by, macroscopic properties of the crystal.

6   Need dispositions have a causal basis?

6.1   Intrinsicality

One of our premises has been that dispositions must have a causal basis. If an object b has the disposition D to manifest M in circumstances T, the causal basis of D in b consists of those intrinsic properties of b, which contribute causally to bringing about M. It would be a mistake to define the causal basis, as Prior, Pargetter and Jackson (1982), 'to be the property or property-complex of the object that, together with [...] [the triggering condition, M.K.] is the causally operative sufficient condition for the manifestation' (Prior et al. 1982: 251). As Molnar (2003: 129) points out, the cause of the manifestation of a disposition is not always exclusively composed of the triggering condition and intrinsic properties of the object to which the disposition is attributed. In many cases, part of the cause of the manifestation M belongs to the circumstances.[12] A match has the disposition to light, but its property of being covered with a layer of potassium chlorate, is only part of the cause of the match's lighting. Oxygen is another part, which is not an intrinsic property of the match. This refutes the claim that every disposition has a causal base *as it is defined by Prior, Pargetter, and Jackson.* But it doesn't refute this claim according to our definition of the concept. If the cause of the manifestation is 'a mixture of circumstances, including some that are intrinsic to the bearer of the disposition and some that are extrinsic to the bearer' (Molnar 2003: 129), there is a causal basis in our sense: the set of intrinsic properties that contribute to cause the manifestation.

However, one might think there are also more extreme cases: dispositions that do not even have a causal basis in our sense. An object b with disposition D, such that *none* of its intrinsic properties contributes causally to a manifestation M of D, would indeed refute our thesis of the necessity of a causal basis.

---

[12]McKitrick's (2003b) 'extrinsic dispositions' belong to this category: the causes of their manifestations are in part extrinsic to the object to which the disposition is ascribed.

Consider the famous chess playing machine built in 1769 to amuse the Austrian Queen Maria Theresa. The machine seemed to have the disposition of playing chess, but in fact it was a dwarf, a human chess player hidden in the machine, who produced the moves on the chess board. Had this machine the disposition to play chess? Whether it is correct or incorrect to attribute this disposition to the robot depends on how exactly one conceives it. Either the robot is taken to be the mechanical device excluding the dwarf. Then it is certainly wrong to attribute any chess playing dispositions to it. However, if the robot is taken to include the dwarf as a part, the attribution is correct. Both possibilities are compatible with our thesis. If the dwarf is no part of the robot, no intrinsic property of the robot contributes to cause the chess moves. But then the robot has no chess playing dispositions in the first place, and there is no reason to postulate any powerful property underlying such dispositions. If the dwarf is taken to be part of the robot, it has chess playing dispositions. But the so conceived robot has intrinsic properties (which are really properties of the dwarf) that contribute causally to the chess moves.

Placebos also seem to have dispositions without any causal basis. Placebo pills can be powerful remedies, yet by definition of a placebo, none of their intrinsic properties contribute anything to cure the patient. Therefore, the pill's disposition to cure the patient does not have any causal basis in our sense. The cases of the placebo pill and of the chess machine share the same structure. To the extent that no intrinsic property of the pill contributes causally to curing the patient it is simply wrong to attribute a disposition to cure to *the pill*. Indeed, it is not really the placebo pill that has the disposition of curing, but the fact that the doctor prescribes it. The causal basis of the curing is in the doctor's act, which cures via the patient's belief that she will be cured.

6.2   Masking

Prior, Pargetter, and Jackson's definition of the notion of a causal basis has another important defect. They take the causal basis of a disposition D to be 'the causally operative sufficient condition for the manifestation' (Prior et al. 1982: 251), at least in the case of what they call 'surefire' dispositions, whose manifestations are produced according to deterministic laws. Let us then disregard here irreducibly probabilistic dispositions, such as radioactive atoms' disposition to decay spontaneously.

It is well known that dispositions can be 'masked' (Johnston 1992; Molnar 1999) or counteracted by an 'antidote' (Bird 1998) without thereby ceasing to exist. 'If we carefully package a fragile vase, thereby masking its fragility' (Molnar 2003: 130) the vase will not break even in circumstances that normally trigger this manifestation of its fragility, such as falling from high on hard ground. The packaged vase is fragile although its properties are not *sufficient* for its breaking when it falls from high on hard ground. Similarly, the poison does not manifest its toxicity if an antidote has also been ingested. Masks and antidotes show that the possession of a disposition is not sufficient to guarantee a manifestation in its triggering condition. Therefore, as Molnar points out, no disposition that can be masked (or for which there exist antidotes) has a causal base *in Prior, Pargetter, and Jackson's sense*.

However, the vase's fragility and the poison's toxicity both have a causal basis *in our sense*. If the vase is fragile, it must be *possible* that its fragility is unmasked. The causal basis is defined with respect to such (normal) circumstances in which the disposition is triggered and manifests itself. The causal basis consists of those intrinsic properties of the object that contribute to the manifestations in these normal circumstances. There cannot be any disposition that is *necessarily* masked (or necessarily accompanied by an antidote): if property P is necessarily accompanied by a mask that prevents manifestation M in triggering circumstances T, then objects having P do *not* have the disposition to M in situations of type T.

6.3   The 'missing base' of fundamental dispositions

Several authors have argued that dispositions of fundamental elementary particles may be 'bare' or 'ungrounded'. This can mean different things. Blackburn (1990: 64) suggests that fundamental dispositions may lack 'categorical ground'. Ellis (2001: 114f) argues that powerful properties do not need 'categorical bases'. McKitrick (2003a: 355) claims that they may be 'bare dispositions', which she defines as dispositions that have no 'causal basis' distinct from themselves. For Molnar, 'no causal bases can be found at all' (Molnar 1999: 9; 2003: 132) for dispositions of simple subatomic particles. Mumford (2006) argues that basic dispositions may be 'ungrounded', i.e. not grounded in some property other than themselves.

Although these claims all express the idea that there are at least some dispositions that are ontologically fundamental, they are not otherwise equivalent. Molnar's argument for his thesis that some dispositions have no basis shows that what he means by a 'basis' is a reduction base. 'Traditionally, to find causal bases we look to the relations between the powers of a whole and the powers and other properties of its parts' (Molnar 2003: 131). Similarly, when Mumford (2006: 471) argues that powerful properties of fundamental properties refute the thesis of 'global groundedness', according to which 'every powerful property is grounded in some property other than itself', he defines 'ground' as meaning 'reduction base': 'The grounds of a powerful property can be found only among the lower-level components or properties of that of which it is a property' (Mumford 2006: 477).

What is distinctive about fundamental particles is indeed their lack of a 'base' for micro-reducing their properties. Such properties of elementary particles as mass, charge, or spin, are the objects of a fundamental scientific theory. In the present state of science, there is no more fundamental science to which that theory might be reduced. Does the absence of any 'reduction base' to which the dispositions of elementary particles might be reduced imply that they do not have any 'categorical ground'? Answering this question presupposes clarifying the conceptual distinction between the dispositional and the categorical. This is a hotly debated issue. Most authors take it to be an ontological distinction between two sorts of properties. However, a minority interprets it as a distinction between two types of concepts or predicates.[13] According to this interpretation, one can conceive of a given property in two ways: either categorically, in terms of what it is, or dispositionally, in terms of which dispositions it gives its bearer. I will adopt the following version of the view that the distinction is conceptual, although I cannot defend it here.[14] A concept or predicate C is dispositional if the truth of the proposition 'b is C', for some object b, entails, a priori (by virtue of the meaning of C), a counterfactual conditional of the form: 'if b was in triggering condition T then ceteris paribus it would produce manifestation M'. A predicate is categorical if and only if it does not, by its very meaning, a priori entail such a counterfactual linking a triggering condition to a manifestation.

In this framework, the relation between a powerful property and the dispositions it gives its bearer can be expressed in the following way. The powerful property, which is the causal basis of (one or several) dispositions to M_i in conditions T_i, appears as categorical if it is conceived as an intrinsic property, and as dispositional if it is conceived indirectly, through the dispositions to M_i in T_i. The charge of an elementary particle is a powerful property, which is the causal basis of its various dispositions to manifest M_i in conditions T_i. This powerful property can be conceived in a dispositional way because it is part of its essence to give its bearer dispositions to M_i in conditions T_i. However, it can also be conceived as intrinsic and categorical.

Does the fact that the dispositions of elementary particles cannot, for lack of parts or internal structure, be micro-reduced in terms of properties of their parts, imply that they do not have any 'causal basis'? It does not: such particles have powerful properties that are the causal basis of the manifestations of these dispositions. McKitrick (2003a: 353) expresses a similar idea by saying that dispositions may be 'their own causal basis'. 'If fragility turns out to be causally relevant to breaking, then fragility is its own causal basis.' However, there is an air of paradox about McKitrick's expression: as a disposition, the fact that vase b is fragile is a priori linked to the counterfactual that if b fell from high on hard ground then, ceteris paribus, b would break. However, it can only be discovered *empirically* whether the vase has such a property as fragility that contributes causally to its breaking in appropriate circumstances. What causes what can only be discovered empirically. This apparent paradox can be avoided by distinguishing, as I have suggested, between the disposition expressed by the counterfactual 'if the vase fell it would break' and the powerful property fragility, which is the causal basis of the disposition.

---

[13] Shoemaker (1980), Mellor (2000).

[14] For a defense, see Kistler (2007).

7   Are there irreducibly multi-track dispositions?

My argument for the distinction between powerful properties and dispositions has made heavy use of the fact that many dispositions are multi-track.
However, the existence of irreducibly multi-track powerful properties has been
questioned by Bird (2007) who argues that dispositions that appear to be
multi-track are always equivalent to conjunctions of single-track dispositions;
furthermore, they are never fundamental. His argument rests on a distinction
between 'pure' and 'impure' dispositions. A pure disposition P can be characterized by a single conditional linking a triggering condition T to a manifestation M. If object b with pure disposition D were in triggering condition T, it
would manifest M.

> (Db and Tb) $\square\!\!\rightarrow$ Mb

What appear to be 'multi-track' dispositions are 'impure' insofar as they
cannot be so characterized. Impure dispositions are linked to many conditionals: There is a whole set of pairs of triggering conditions T_i and manifestations
M_i, such that an object b with an impure disposition I would manifest M_i if it
were in condition T_i.

> For all i [(Ib and T_ib) $\square\!\!\rightarrow$ M_ib].

Bird holds the following two theses.

> (T1) All impure dispositions are conjunctions of pure dispositions,
> and
> (T2) All impure dispositions are non-fundamental. (Bird 2007a:
> 22)

Take our example of conductivity σ. It seems to be an impure disposition,
insofar as it is characterized by a whole set of counterfactual conditionals. For
each value of electric field E_i, the current density is $J_i = \sigma E_i$. Application of (T1)
yields the claim that attributing σ to an object is equivalent to attributing to it
the conjunction of dispositions:

> (CD) For all i [(E_i) $\square\!\!\rightarrow$ J_i].

Thesis T1 quite naturally leads to thesis T2 according to which all impure
dispositions are non-fundamental. Indeed, Bird argues, there cannot be an
impure fundamental disposition, because 1) given thesis T1, impure dispositions are conjunctions of pure dispositions, and 2) conjunction of dispositions
cannot be fundamental, since a conjunction is less fundamental than its conjuncts.

Here are two arguments against Bird's first thesis. The first uses Bird's (implicit) distinction between two types of 'impure' dispositions:

Type A: properties such as electrical conductivity, which figure in several deterministic functional laws relating variables with infinitely many values.

Type B: propensities, which are dispositions with probabilistic manifestations, such as dispositions for different types of radioactive decay.

Bird allows, although somewhat tentatively, that properties of type B may be irreducibly multi-track. Let us suppose that particles of type Z, when triggered by condition T, either decay in the A-cascade or in the B-cascade, each with its intrinsic probability. For such a particle x, if x is triggered by T, then either, with probability α, it decays in the A-cascade, or with probability β, it decays in the B-cascade. Is this disposition equivalent to the conjunction of the disposition to decay in the A-cascade and the disposition to decay in the B-cascade? It is not, because there is no ground for distinguishing two dispositions here: necessarily, something has one if and only if it has the other. In such a case, Bird admits, 'we would not wish to ascribe different intrinsic dispositions', and concludes that 'such cases are best assimilated to propensities' (Bird 2007a: 23, fn18). Contrary to (T1), propensities seem to be impure dispositions that are not equivalent to conjunctions of pure dispositions.[15] The propensity of the Z particle seems to be equivalent to a powerful property and distinct both from the (pure) disposition to decay in the A-cascade and from the (pure) disposition to decay in the B-cascade. Now, if our distinction is needed to account for propensities, it is more ontologically parsimonious to make the hypothesis that it applies to all multi-track dispositions.

My second argument bears directly on the controversial powerful properties of type A. The reasons for postulating powerful properties are the same as the reasons for postulating theoretical properties, over and above the conjunction of their test sentences, which are conditionals associating triggering conditions to manifestations.

Take any law, such as $J = \sigma E$, that expresses in a single formula the dependence of the values of J, on circumstances E. According to (T1), attributing σ to an object is equivalent to attributing to it the conjunction (CD). However, as I have argued above, the explanatory power of a theoretical property such as σ goes beyond the conjunction of its test conditionals. The attribution of σ to an object is the first step towards a unifying explanation of all these counterfactuals. Rather than just abbreviating them, it indicates their common truthmaker.

A simple fact that the attribution of σ can explain, and the mere conjunction of conditionals cannot, is why there could not be an object that has just *part* of the conjuncts of (CD). If the conjunction were a brute fact, this would be possible. However, if all the conjuncts are consequences of the object's having the powerful property σ, one of these conditionals can only be true of a given object if the whole set of them is also true of it.

---

[15]Fundamental propensities also refute (T2).

There is also a more scientifically fruitful sense in which the content of the attribution of a theoretical property goes beyond the mere attribution of the conjunction of its triggering-manifestation conditionals. Compare the conditionals associated with the electrical conductivity σ and with the thermal conductivity λ.

(CD-σ) For all i [(E$_i$) $\Box\!\!\rightarrow$ J$_i$].

(CD-λ) For all i [(ΔT$_i$) $\Box\!\!\rightarrow$ Q$_i$].

Only if one postulates the existence of properties σ underlying the conjuncts of (CD-σ) and λ underlying the conjuncts of (CD-λ), does it become possible to state a link between them: the Wiedemann–Franz law, and to explain that link: by reducing the law.

According to Bird's (T2), there are no fundamental impure dispositions. However, the same reasoning applies to fundamental dispositions as to non-fundamental dispositions. Even for fundamental properties, the powerful property is more fundamental than 1) each individual disposition it gives its bearer, and 2) the conjunction of these dispositions. The reason is that it explains and unifies all these dispositions. The difference between fundamental and non-fundamental depends on whether or not a powerful property has a reduction base. However, this difference is independent of the relation between powerful properties and dispositions.

8   Conclusion

I have argued that multi-track dispositions can best be understood if we distinguish the *powerful property* that makes it *one* multi-track disposition, from the various *dispositions* that correspond to its different manifestations in different circumstances. Powerful properties are real properties insofar as they are efficacious. Indeed, we have defined them as those intrinsic properties of an object possessing a multi-track disposition, which contribute causally to all of its manifestations. The reasons for taking powerful properties to be real are the same as for other theoretical properties. They provide explanatory simplification and unification, insofar as they constitute the common ground of the various dispositions associated with a multi-track disposition. The fact that elasticity underlies, among other manifestations, retracting, expanding and bouncing of rubber balls in different circumstances, explains why it is no accident that objects having one of these dispositions in general also have the others. Much confusion in this debate results from the failure to distinguish the causal basis from the reduction basis. Indeed, the fact that fundamental dispositions have no reduction base is no reason to deny that they have a causal basis. Insofar as they have typical manifestations, all dispositions have causal bases: powerful properties. Whether a powerful property is fundamental or not, and in the latter case, whether it is reducible or not, is independent of the question of its efficacy in contributing to bring about the manifestations of the dispositions it underlies.[16]

---

[16]I thank my audiences in Nottingham and Melbourne, as well as D. H. Mellor, Reinaldo Bernal, Alfredo Paternoster and an anonymous referee, for helpful criticisms.

FOUR THEORIES OF PURE DISPOSITIONS

William A. Bauer

1 Pure Dispositions and the Problem of Being

According to one theory of dispositions, any dispositional property requires a causal basis consisting of one or more categorical properties (e.g., see Prior et al. 1982). However, several metaphysicians and philosophers of science have recently claimed that pure dispositions (or, pure powers) are either actual or at least metaphysically possible.[1] Bird (2007a), Ellis (2001), McKitrick (2003a), Molnar (2003), and Mumford (2006) all accept the following thesis or something akin:

> Pure Dispositions Thesis: It is metaphysically possible that there is some type of dispositional property of which any instance, F, does *not* have a distinct causal basis for its manifestation, where a causal basis consists of some instance of either another dispositional property or property-complex, or a categorical property or property-complex, or combination thereof.[2]

The causal basis is a property or property-complex, either constituted by dispositional or categorical properties, that is causally relevant to the manifestation of a disposition when appropriately triggered in the right circumstances.

---

[1]Moreover, the theory of dispositional essentialism (Bird 2007a; Mumford 2004) requires that all of the fundamental properties are pure dispositions, in contrast to the claim that all of the fundamental properties are categorical (Armstrong 1997, 2004).

[2]Two notes: First, the Pure Dispositions Thesis says that F does not have a *distinct* causal basis, since F might be its own causal basis, as McKitrick (2003a) argues. Section 2 addresses this further. Second, if dispositional essentialism is true, then pure dispositions are metaphysically *necessary*, so of course possible; however, attending to the mere metaphysical possibility of pure dispositions is sufficient for the purpose of this paper.

For example, a vase possesses the dispositional property of fragility. The disposition may be triggered by a hammer striking the vase, where it is some set of micro-structural properties of the vase that constitutes the causal basis of that particular instance of fragility.[3]

Importantly, besides being causally relevant to the manifestation of a given disposition, the causal basis also plausibly accounts for the disposition's continuous existence or being. So the grounds and causal basis of a disposition are, it appears, identical. When the disposition is not manifesting, the causal basis somehow anchors or grounds the being of the disposition. For example, the continuous existence of fragility is grounded by micro-structural properties of the fragile object. (The grounding relation is further specified in Section 2, where I will challenge the assumption that the causal basis is identical to the grounds of a disposition.)

Given that pure dispositions have no causal bases, and that causal bases typically ground the being of dispositions, the Pure Dispositions Thesis raises the question of how pure dispositions continuously exist. This paper aims to investigate and solve this problem. Here is the problem more precisely stated:

> Problem of Being: For any instance of a pure disposition, F: assuming that F need not manifest continuously, and assuming that there is no property or property-complex distinct from F that constitutes a causal basis that grounds F, there is nothing to ground the continuous existence (or, being) of F.

Another way of putting the problem: In what does the being of a pure disposition consist, apart from its possible manifestations? Or, when a pure disposition at space-time location l is not manifesting, *what* is at l? Or, as Psillos (2006b) asks: What does a pure disposition do when not manifesting? This problem is at the heart of many worries about pure dispositions, and systems of properties built up from them.[4] Extensive arguments have been given for the possibility and actuality of pure dispositions.[5] However, the nature of the continuous existence or being of pure dispositions appears to have received little attention.[6]

---

[3] Examples of pure dispositions typically offered are the properties of fundamental particles, such as mass, charge and spin. I sometimes use these examples, though it may be that some further properties are more fundamental. Moreover, so far as the Pure Dispositions Thesis says, there may be emergent, higher-order pure dispositions.

[4] See especially Heil (2003: 97–110), Mumford (2006), and Psillos (2006b). Although he supports the Pure Dispositions Thesis, Mumford (2006: 485) explicitly raises the question of the being of pure dispositions.

[5] See McKitrick (2003a) for an argument for the possibility of pure dispositions based primarily on metaphysical considerations, and Molnar (2003: 125–42) and Mumford (2006) for arguments for the actuality of pure dispositions based jointly on metaphysical and empirical considerations.

[6] Two exceptions are Handfield (2008) and Mumford (2006). Mumford (2006: 485) suggests that pure dispositions are self-grounded, and Handfield (2008) suggests that pure dispositions are either self-grounded or globally grounded. Both theories are discussed below. Both Mumford (2006) and Handfield (2008) suppose that McKitrick's idea (2003a) that a disposition may be its own

One may think that the being of pure dispositions needs no explanation, especially if one thinks that pure dispositions form the ground floor of reality, as for example dispositional essentialism maintains.[7] Supposing that explanations must stop somewhere, one might simply accept the continuous existence of pure dispositions without explanation, just as one might accept the continuous existence of some other class of fundamental entities posited on some other metaphysic.

However, I have three responses to any 'no explanation' response to the Problem of Being. First, the same puzzling metaphysical questions previously raised about pure dispositions remain, e.g., about what pure dispositions do when they are not manifesting. Second, regardless of whatever considerations in favor of a 'no explanation' view one might offer, we should not consider the Problem of Being alleviated without at least canvassing various possible answers to it. There may be a plausible metaphysical explanation not yet identified or developed in sufficient detail, e.g., a solution that explains the being of pure dispositions by reference to conditions extrinsic to the objects that bear them. Third, given the modal nature of pure dispositions it seems there *must* be some explanation of their being when not manifesting, even if we judge that there are good reasons to posit them absent an explanation of their being; e.g., we may have sufficient reason to posit pure dispositions because of their explanatory power, yet this does not explain the continuous existence of pure dispositions. Moreover, if there is no explanation of the being of pure dispositions, then such entities would just seem to 'hang' ontologically on nothing, akin to how psychological dispositions do on Ryle's (1949) account of mind. Even if pure dispositions do 'hang' in this way, some explanation of what this amounts to is in order. Is it a causal process of some sort? Is it self-grounding, and if so what does this amount to? In light of these considerations, I suggest that taking the 'no explanation' stance is only epistemically permissible

---

causal basis naturally suggests that a disposition is its own grounds as well. I think this is ultimately right (and I offer an account to support it in Section 7), but this assumption is also questioned below (Section 2) by giving reasons for thinking that grounds and causal bases are different. This opens up several possible grounding options for pure dispositions that deserve evaluation.

[7] Dispositional essentialists such as Bird (2007a) and Mumford (2004) maintain that all the fundamental properties are purely dispositional, but this does not necessarily mean that they think no explanation is required in response to the Problem of Being. Ellis (2001), who accepts the Pure Dispositions Thesis but does not think *all* of the fundamental properties are essentially dispositional (he argues that fundamental spatiotemporal and numerical relations are categorical), argues for the 'no explanation' response to the Problem of Being. Ellis (2001: 114, and especially 2001: 139–40, fn 12) maintains that the existence of higher-level entities plausibly depends on increasingly lower-level causal powers (dispositions), ultimately bottoming out in purely dispositional properties at the fundamental level of reality. Since we must ultimately posit fundamental pure dispositions (barring infinite levels), on this view, their continuous existence needs no explanation. Although it may be true that we must posit them, it is not clear why they must be posited with no explanation as to their continuous existence, for reasons given in the next paragraph. Furthermore, Ellis seems to assume that pure dispositions are intrinsic properties; however, if they are extrinsic they will necessarily have some grounds (though not necessarily causal bases, per my arguments in Section 2), and these grounds will figure in an explanation of their being.

once we have reason to reject various theories of the being of pure disposi-tions. Hence, we should seek such a theory.[8]

In this essay, I evaluate four theories that aim to solve the Problem of Being. I will proceed as follows. In Section 2, I establish criteria for evaluating the theories, and I argue for two assumptions that will hold throughout the rest of the essay. In Sections 3 through 5, I present three theories that I argue do not satisfactorily answer the Problem of Being. I take none of the reasons I give against the three theories to be absolutely conclusive, but only to count against them; certainly each theory could use more extensive development, but I sketch enough of each sufficient for an initial evaluation. In Section 6, I argue that the most viable theory, i.e., the one that fully satisfies the two criteria established in Section 2 and has the fewest additional problems, maintains that a pure disposition grounds itself: a pure disposition is the grounds of its own being. This basic idea has been advanced by others, for example Mumford (2006) advocates the self-grounding of pure dispositions, but not much has been said about *how* pure dispositions are self-grounded or what exactly self-grounding amounts to. In Section 7, I advance and develop an explanation of how pure dispositions are self-grounded, i.e., of what it means for a pure disposition to be self-grounded.

2   Criteria and Assumptions

As the Problem of Being indicates, for any pure disposition, F, any satisfactory theory of pure dispositions should satisfy the following criteria:

> Criterion 1: The theory explains the continuous existence or being of F.

> Criterion 2: The theory does not invoke additional properties that con-stitute a casual basis for F's manifestation.

I will reference these criteria in evaluating the four theories, though other con-siderations will also be discussed regarding each theory. Additionally, two as-sumptions will hold throughout the evaluation of the theories. For any pure disposition, F:

> Assumption 1: F may serve as its own causal basis.

> Assumption 2: There is a metaphysical distinction between the *grounds of F* and the *causal basis of F*.

---

[8]It is worth noting that it does not seem sufficient as an answer to the Problem of Being to of-fer a characterization or analysis of dispositions (and pure dispositions), such as the directedness theory of dispositions (e.g., Molnar 2003) or some version of the conditional analysis of disposi-tions. These theories only raise more questions about the being of pure dispositions: e.g., on the directedness theory, what does it mean for F to be in a state of directedness when not manifesting, and what is so directed? And, on the conditional analysis, when not subjected to its manifestation conditions stipulated in the antecedent of the condition, what is the nature of F's being?

Assumption 1 is McKitrick's proposal (2003a): a pure disposition (or, 'bare disposition' in her terminology) has no causal basis either in further dispositions or categorical properties. When a manifestation occurs, it is F itself (and not some subvenient property of F) that is activated by an external stimulus, triggering F to manifest. Hence, F need not have a *distinct* causal basis. This is an important component of the theory of pure dispositions I advocate below (the fourth theory) since some explanation is needed to explain how or what is activated when F manifests.

However, if a pure disposition is its own causal basis that does not necessarily mean that it grounds itself. For purposes of this paper, the grounds of a property, P, consist of another property or property-complex upon which P ontologically depends for its continuous existence. For example, an instance of a color property, R, is grounded partially in R's object-bearer having the property of extension. Thus, a tennis ball's extension grounds its property of being yellow, but does not *cause* yellow. Moreover, yellow does not necessarily supervene on the shape of the tennis ball; the tennis ball can change shape while retaining the same color. The intuitive idea is that the grounding properties are those that, if eliminated, would result in the immediate (simultaneous) elimination of P as well. (It must be simultaneous otherwise an animal's property of being alive, for example, would be grounded in properties of oxygen, which appears to be a causal, not a grounding, relation.) Also, P can have partial grounds in distinct sets of properties; e.g., P might be grounded partially in properties of its object-bearer, x, and also be grounded partially in properties of some object in x's environment, or property of x's environment. (This is the case with relational properties, such as tallness). In the case of color, a color R seems partially grounded in light-reflection properties of the object bearing R and also partially grounded in the object's extension.

Given this understanding of grounding, the distinction in Assumption 2 is that there is a difference between the basis for F's continuous existence (i.e., ontological grounds) and the basis for F's manifesting given an appropriate stimulus (i.e., causal basis). I will argue that the causal basis and grounds of a disposition may come apart in two kinds of cases—extrinsic and intrinsic.[9]

In the intrinsic case, it is possible for F to ontologically depend on and therefore be grounded in intrinsic properties of the object x that bears F, but that are *not* part of the causal basis of F. The suggestion is that necessary co-instantiation of two properties, P1 and P2, by x, is a kind of ontological grounding or dependence between P1 and P2. For instance, suppose it is necessary for the disposition mass to exist that it bundles with charge and spin, so that the being of mass is grounded in the bundle *charge-spin-mass* (there's no possible world in which you find the disposition mass that you do not also find mass bundled with the dispositions spin and charge, thus forming an electron

---

[9]Handfield (2008: 298) distinguishes between the supervenience base of a disposition and the causal basis for the manifestation of a disposition, akin to my distinction between grounds and causal basis. I address this below.

for example). Yet, it is not obvious that the charge and spin in a particular
electron bundle are causally relevant to the manifestation of mass, i.e., that
the charge and spin form the causal basis for mass' manifestation. It does not
seem as if the charge-spin component of the bundle needs to receive the stimulus in order for mass to manifest. In this example, mass may be its own causal
basis yet it is ontologically grounded in—i.e., ontologically dependent on—its
bundle partners charge and spin, for it would immediately cease to exist without them. Thus, the grounds of F and causal basis of F are not necessarily
identical.

In the extrinsic case, it is possible that F, of object *a*, is partially ontologically grounded in properties of some object *b* that does not bear F, yet *b* is not
obviously causally relevant to (by which I mean that it forms part of the causal
basis for) the manifestation of F (given an appropriate stimulus). For example,
take the extrinsic disposition *vulnerability*—i.e., capable of being damaged if
attacked (this is an example of an extrinsic disposition discussed by McKitrick
(2003b)). For some object *x*, *x*'s being vulnerable depends on whether *x* is an
environment that affords protection to *x*. When *x* is *not* in the protective environment, *x* is vulnerable. When *x* is in a protective environment, *x* is not
vulnerable. Thus, vulnerability is an extrinsic disposition. But it is not obvious
that *x* being located in a non-protective environment (and so being vulnerable) is part of the causal basis for the manifestation of *x*'s becoming damaged
(where being damaged is the manifestation of vulnerability); rather it seems
like it is solely some properties of *x* itself that constitutes the causal basis for *x*'s
vulnerability, even though *x*'s having the disposition of vulnerability requires
extrinsic (environmental) factors, i.e., the lack of appropriate protection. Thus,
the grounds of vulnerability and the causal basis of vulnerability are distinct.

Given Assumptions 1 and 2, the question I am pursuing is: how can a pure
disposition, F, be ontologically grounded, but nonetheless remain a pure disposition in that it does not have a distinct causal basis? The metaphysical distinction between the grounds and causal basis of a disposition (i.e., Assumption 2) opens up several possible theories of the grounding of a pure disposition. This is because we can look for ways to explain the grounds of F's being without necessarily positing a causal basis for F, thus violating Criterion
2. However, what I will ultimately suggest in the self-grounded theory of pure
dispositions (Section 6) is that F is indeed both its own causal basis *and* its
own grounds for being. I advance this as a metaphysically necessary claim, yet
other theories are at least logically possible, and so must be ruled out before
we can infer that F is self-grounded.

Here is the structure of the overall argument of this paper:

(1) Rule out these possible grounding theories as solutions to the problem
     of F's continuous existence:

     (i) F is extrinsically grounded in all properties (i.e., global grounding);

   (ii) F is extrinsically grounded in properties of the World, taken as a whole ontologically prior to the part-objects of the World;

   (iii) Supposing that F's object-bearer x is substantial, x grounds F.

(2) If (i), (ii), and (iii) are ruled out, then Assumption 1 implies that F is self-grounded.

(3) Thus, Assumption 1 implies that F is self-grounded. [(1), (2)]

(4) If F is self-grounded, then the intrinsic nature of F accounts for F's continuous existence.

(5) Thus, the intrinsic nature of F accounts for F's continuous existence. [(3), (4)]

This argument relies on the plausibility of ruling out other grounding options, i.e., premise (1), the task to which I now turn in Sections 3 through 5. Then, Sections 6 and 7 develop premises (2) and (4).

3   The Global Theory

Handfield (2008: 298) distinguishes between the 'supervenience base' and the 'causal base' of a disposition, similar to the distinction between grounds and causal basis in Assumption 2. He then suggests that though pure dispositions lack causal bases, they 'do have supervenience bases, but that they represent a degenerate case: their base includes every possible property, including extrinsic properties' (2008: 304).[10] If the supervenience base of F is something like the grounds of F, then the theory being suggested (though not necessarily endorsed) by Handfield is this:

   (Global) The ontological ground of F consists of every possible property, including extrinsic properties.

By 'every possible property' Handfield seems to mean (he does not say explicitly) every property in the actual world, not also properties of possible worlds. (Besides, it is a mystery how possible properties would ground actual properties.) So, the ground of F consists of the entire set of actual properties in a given world. In characterizing the supervenience base of a pure disposition as degenerate, I take him to simply mean that the base deviates from the common assumption (e.g., Mumford 2006) that the grounding property (or property-complex) of a disposition must be intrinsic to the disposition's bearer.

If (Global) is correct, then the set of global properties grounds F while not being a causal basis for F. Perhaps the best example of something approaching the content of (Global) is Mach's Principle, which maintains that the mass of an

---

[10]Terminological note: Handfield (2008: 304) calls this the 'global hypothesis', and he uses the term 'bare dispositions' instead of pure dispositions (following McKitrick (2003a)).

object, x, is determined by the total distribution of mass and energy in the rest of the system of which x is a member. Thus, according to this principle mass is an extrinsic property. This principle is limited as an example of (Global), according to which *every* pure disposition, not just mass, is grounded in all other properties. But Mach's Principle at least serves as an intuitive analogy.

On (Global), the continuous existence of F is explained by appealing to all other properties, thus invoking the idea of the interconnectedness of all being. We will also see this idea in theories below. Let us assume Criterion 1 is satisfied. What about Criterion 2? If it is true that the global grounds do not play a role in F's manifestations, then Criterion 2 is satisfied. It is not a causal basis for F, one might argue, because all the properties in the supervenience base are extrinsic relative to the object bearing F. How could they be causally relevant to F's manifesting? How could the property of the moon looking beautiful to Galileo be causally relevant to some pure disposition's manifestation event? However, if even *one* is causally relevant, then part of the grounds of F is also a causal basis for F, and so F is not pure. Most of the global properties are going to be extrinsic relative to the object bearing F. But some will not, for some might be intrinsic properties of the object bearing F. But that may not be a problem for (Global) if those properties are simply other pure dispositions (as is plausible for the subatomic particles). Thus, if the global grounding properties are not part of the causal basis of F, yet they do form ontological grounds for F, then we have a solution to the Problem of Being.

Nonetheless, one might argue that the global grounding base *does* also form a causal basis for F somehow. This might be true in two ways: first, perhaps *some* of all the properties of the world may be causally relevant to F's manifesting; second, perhaps all of the properties form a causal basis for F's manifesting in some 'degenerate' sense, in the same sense that they all form a 'degenerate' supervenience base or grounds. (Note that if it really is a supervenience base as Hanfield thinks, then some property or property-complex G in the base must co-vary with F, and since G is in the base of the supervenience relation, it looks like it will be causally relevant.) We might say that F has a 'hidden' or 'subdued' causal basis. It may be that F manifests in some circumstance even though F is not directly stimulated, because some or all of the global properties are stimulated.[11]

Setting aside the question of whether the global base just ontologically grounds F or also serves as a causal basis for F, (Global) is unappealing for two reasons. First, the *truth-maker* for the proposition 'F continuously exists' (or

---

[11] It is worth clarifying that some properties of the world besides F will be causally relevant to F's manifesting if (Global) is true, simply because the stimulus of F will be causally relevant. But typically the triggering or stimulus property is *not* considered part of a causal basis for any disposition. I assumed that was so in advancing the first possibility for what the causal basis might be, to the effect that the entire global grounds could not be the causal basis yet the causal basis could be some sub-set of the global properties not including the stimulus.

'F is grounded') is every property of the world.[12] This is problematic on two counts: first, it seems like an unnecessarily large truth-maker, and second, one could reasonably claim that every property's truth-maker is the set of all other properties of the world. It may be true in a weak sense that all other properties make true the proposition 'F continuously exists', or 'the state of all the properties in the entire world is such that F is grounded'. This reflects Armstrong's (2004: 19) observation that the whole world is the 'least discerning' (because of its non-specificity) and 'most promiscuous' (because it makes every truth true) truth-maker. But this is not very informative; what would be informative is a 'minimal' truth-maker (Armstrong 2004: 19) for F.

The second consideration against (Global) is that not *every* other property seems relevant to whether F continuously exists at a specific location. If spin is a pure disposition, then on (Global) the property of earth being the fourth planet from the sun forms part of the grounds for an electron on Venus having spin ½. This is wildly counter-intuitive. Perhaps some more precisely defined set of global properties forms the grounds of F, such as all other sparse properties, or perhaps just properties of the world as a whole. But, to entertain that conception of F's grounds is to entertain a different theory than (Global).

In sum, even if (Global) fares decently on Criteria 1 and 2, other considerations lead to a tentative rejection of this theory of the grounding of pure dispositions.

4 The Monistic Theory

In this section I consider two versions of monism to account for the continuous existence of pure dispositions. Monism is the idea that the World (i.e., the entire cosmos or universe) is a genuine object, not just a collection of smaller objects (the World is tantamount to the One discussed in Plato's *Parmenides*). The two versions of monism I will consider are priority monism and existence monism, a distinction made by Schaffer (2007, 2010b).

Existence monism maintains that the World is the *only* object. Horgan and Potrč (2008) call it the Blobject. On their view, confluences of properties of the Blobject form what we call 'objects', e.g., atoms, cells, diamonds, and trees. By contrast, priority monism as defended by Schaffer (2010b) holds that the World—again, the whole cosmos or universe—is ontologically prior to all the smaller objects in the world, but all those smaller parts of the World (the atoms, cells, diamonds, trees, etc.) are legitimate objects in their own right. The World

---

[12]This objection should not be confused with the objection that the manifestation conditional associated with F is made true by every property in the world. Although this too may seem counter-intuitive, this objection would assume that all the properties of the world are somehow causally relevant to F's manifesting, and thus form a causal basis for F, which would violate Criterion 2. (The associated manifestation conditional is 'the state of the entire world is such that, were x exposed to the characteristic stimulus S, it would yield the characteristic manifestation M', Handfield 2008: 304.)

has primary being and its part-objects have derivative being; i.e., the World has ontological priority over the part-objects.

Assume priority monism is true. Priority monism entails that some property or property-complex of the World partially grounds all of its part-objects and by extension also the properties of those part-objects. Thus, any pure disposition F, borne by a fundamental object *a*, is partially grounded in a property or property-complex of the World. So the theory is this:

> (World) F is grounded in the World, the whole that is ontologically prior to all the part-objects in the cosmos.

This theory has an important implication. Properties, including dispositions, that are ontologically dependent on properties of objects (or environments) *other* than their bearers are extrinsic properties, and thus are grounded extrinsically. Thus, if priority monism is true, then pure dispositions are extrinsic.

(World) and (Global) both invoke the idea that F is grounded in properties external to the object bearing F. On both, F qualifies as extrinsic because F's being is not solely dependent on properties intrinsic to the object bearing F. But, on (World) the extrinsic grounding base is less robust in the sense that it only invokes properties of (or a single property of) the World as part of the grounds of F, whereas (Global) invokes all other properties beyond F. Compared to the truth-maker problem associated with (Global), (World) yields a more plausible truth-maker for 'F is grounded', since only the World (i.e., some property or set of properties of the World) preceding the part-objects is referenced, not *all* properties. So, per Occam's razor, (World) should be favored over (Global).[13] Furthermore, F does not depend on *non*-fundamental properties as it does on (Global); F only depends on fundamental properties of the World. That is significant if pure dispositions must be fundamental properties, since it is counter-intuitive for fundamental properties to depend ontologically on non-fundamental properties.

Criterion 1 is satisfied since the World grounds everything, including all properties. As may be true with (Global), (World) does not necessarily violate Criterion 2, since the grounding properties of the World that account for F's being are not necessarily part of the causal basis of F's manifesting. Note that considerations discussed in Section 3, pertaining to Criterion 2 relative to (Global), are also relevant to Criterion 2 relative to (World). However, in addition to those considerations, (World) and (Global) both pose the question of whether extrinsic grounding properties form part of the causal basis of F, thus violating Criterion 2. Some comments on that are in order, and these comments are relevant to (Global) as well, but I will couch my discussion of extrinsicness in terms of (World).

On (World), F could still be its own causal basis if the grounding properties in virtue of which F is extrinsic are not part of the causal basis for F's manifestations, given that grounds and causal bases may come apart (Assumption 2).

---

[13] See Schaffer (2010a) for a 'one truth-maker' view.

But *if* the extrinsic grounds form a causal basis for F, one might respond on be-half of (World) that it is intrinsic purity (i.e., intrinsic 'causal base purity') that matters to whether F is a pure disposition. Thus, even if the extrinsic ground-ing properties form a causal basis for F, they at least do not form an *intrinsic* causal basis. But why does this matter? If the extrinsic base is causally relevant to F manifesting, then it should not make a difference between the intrinsic-ness/extrinsicness of the causal basis—a causal basis is a causal basis. Still, it might be urged that if the extrinsic grounds of F form a causal basis, their casual relevance in F's manifestations are only indirect, in a way that the en-vironmental conditions of any dispositional property are causally relevant to what happens in manifestation circumstances. So in this sense, F might have an extrinsic grounds and extrinsic causal base but still be pure intrinsically.

I think there is a good case to be made that the extrinsic grounds of F do not also serve as a causal basis for F, thus making (World) an attractive solu-tion to the Problem of Being. However, a significant problem is that (World) issues from a radically different conception of the nature of the world than most philosophers discussing pure dispositions assume. Although Schaffer (2010b) defends priority monism, the standard assumption is *pluralism*, the view that the world is composed of abundant fundamental objects whereas priority monism says there is only one fundamental object. I will not eval-uate these two competing theories of the number of fundamental objects in this essay. But major proponents of pure dispositions, such as Bird (2007a) and Mumford (2004), assume pluralism. Provided that a theory of the being of pure dispositions should be consistent with pluralism, at least for dialectical reasons, there is reason to tentatively set aside (World).

Another monistic option to account for the being of pure dispositions in-vokes existence monism. As discussed above, existence monism says that there is just one metaphysically genuine object, the Blobject. As an answer to the Problem of Being, we get the following:

(Blobject) F is ontologically grounded in virtue of being instanti-ated by the Blobject.

On this view all pure dispositions are grounded in the Blobject directly. If pure dispositions are fundamental properties, then they are the fundamental prop-erties of the Blobject and their continuous existence is explained by the contin-uing existence of the Blobject. So, Criterion 1 is satisfied. Unlike (Global) and (World), (Blobject) does not invoke extrinsic properties to account for F's be-ing simply because there is nothing extrinsic to the Blobject in virtue of which it has its properties. F is simply instantiated by the Blobject as any typical ob-ject instantiates a property (although on this version of monism there is only one genuine object). If there are no mediating properties (categorical or dis-positional) between the Blobject and F, then no properties besides F itself con-stitute a causal basis for F, and Criterion 2 is satisfied.

I have two responses to (Blobject). First, as with (World), given the dominant pluralistic view of objects it is dialectically inadvisable to assume that the
Blobject exists. It is merely a possibility to account for the being of pure dispositions and warrants further investigation. Second, and more importantly,
(Blobject) raises a question about the nature of the relation between objects
and pure dispositions, assuming that pure dispositions must be borne by objects and cannot 'float free' or potentially exist apart from some object. The
question is this: if the instantiation of F by an object ontologically grounds F,
and that object has the additional property, G, of *instantiating* F, then how is it
that G is not thereby a causal basis for F? The next section will clarify this question, but note that the Blobject shares this potential problem with any ontology that maintains that properties must necessarily be borne or instantiated
by objects.[14]

5   The Object Theory

Suppose that pure dispositions must be instantiated by objects.[15] Perhaps an
object bearing or instantiating F accounts for F's continuous existence without
invoking any further properties of the object that would constitute a causal
basis for F's manifestation. Thus, for example, an electron's instantiation of a
dispositional property token of charge accounts for the continued existence of
that charge token during periods of non-manifestation.

In the dialectic concerning pure dispositions, some implicit assumptions
are often made concerning the nature of objects, besides the fact that it is often
assumed that pure dispositions are necessarily instantiated by objects. Sometimes it is assumed that the objects bearing pure dispositions are just bundles
of tropes. And sometimes it is assumed that the objects bearing pure dispositions are substantial objects (i.e., substances distinct from their properties)
that instantiate pure dispositions and other properties. On the former view, if
some object is just a bundle of pure disposition tropes, then it is not clear how
the object could account for the continuous existence of its pure dispositions
when they are not manifesting, since there is nothing beyond the pure dispositions in the bundle. But on the latter view, the object could account for F's
continuous existence because it is a substantial entity that *has* properties and
the properties of the object depend on the object for their existence.

It is the substantial view of objects that I have in mind in evaluating objects as grounds of the being of pure dispositions. I will argue that objects, in
this sense, cannot straightforwardly account for the being of pure dispositions

---

[14]The fact that (Blobject) is just one way to cash out an object-centered theory of the being of
pure dispositions is another reason, in addition to it being a variation of monism along with priority monism, why I do not consider (Blobject) as one of four distinct theories of pure dispositions
discussed in this essay.

[15]McKitrick (2003a: 254), for example, holds that pure dispositions are dispositions *of* objects,
or dispositions that objects *have*.

without violating their 'purity', and thus if pure dispositions exist, then they are not instantiated by substantial objects. Thus, pure disposition tokens can either 'float free' of objects or are instantiated by objects as bundles of properties, and so some other theory of their ontological grounding is needed.[16] This does not necessarily imply that there are no substantial objects at higher levels of reality, for macro objects may be constituted by smaller objects that are bundles of properties.

Suppose that F is grounded by the object of which it is a property, and that the grounding relation between F and its object, O, is simply F's being instantiated by O. The theory under consideration is thus:

> (Object) F is grounded by O insofar as O is a substantial object that instantiates F.

F is a property instantiated by O (or, exemplified by O) and since O continuously exists, F continuously exists. So an electron's charge disposition continuously exists simply because the electron, as a substantial object that bears that charge token, continuously exists.

Assuming the substantial view of objects and that objects instantiate properties, we can interpret (Object) in two ways. First, properties are *ways* objects are; second, properties *inhere in* objects. The latter idea, as I am using it, is to conceive of objects as discrete substrata onto which properties are 'pinned'. I am primarily concerned with the former view of objects in discussing (Object). This is because on the second view properties would be distinct entities that are separable from their object bearers, and so a problem would be that properties do not necessarily require their object bearers to exist, i.e., property tokens could possibly float free. Thus, in addition to assuming that objects are substantial, in discussing (Object) I will further assume that properties are *ways* objects are, and thereby avoid possible worries about the view that properties inhere in objects. (Perhaps these two views of the relation between properties and substantial objects amount to the same thing; if so, then my critique is not affected.)

As a model for evaluating (Object) I will adopt the view of Lowe (2006), who argues that property tropes or tokens are *ways* objects are; in other words, property tokens are modes of objects. Moreover, Lowe (2006: 27, 75) thinks that property instantiation necessarily depends upon objects: there can be no property-instance that is independent of an object, for property-instances just are properties *of* objects (so no free-floating properties).[17] This implies that

---

[16]It may be that any property, and so a pure disposition too, can float freely without inhering in an object and so does not need an object to be instantiated. I do think this is possible; see Schaffer (2003b) for a defense of this view. I return to this possibility in Section 7.

[17]Although Lowe (2006) does not accept a substratum view of objects, he does include objects as distinct kinds in his Four-Category Ontology. Properties are ways objects are, not things that are pinned on their object-bearers. Heil (2003: 173) also holds that 'properties are ways objects are.'

objects ontologically ground dispositional properties since one of the ways an object might be is to be disposed to manifest a certain way.[18]

Criterion 1 is satisfied on (Object), since O's continuous existence will ensure F's continuous existence. An event may *cause* O to lose F, as an apple may lose its redness due to decay, yet when O has F it is O that is ontologically responsible for F's continuous existence. To maintain the purity of F on this view, and thus satisfy Criterion 2, it must be that O does not constitute a causal basis for F's manifestation. Therein lurks a potential problem for (Object). O will be involved in interactions with other objects, and I suggest that some further property of O must be stimulated in some way so as to bring about the manifestation of F; what this further property of O is will be addressed below. But if some further property of O plays a causal role in F's manifestations, then it looks like F is not its own causal basis, as Criterion 2 requires. So the worry is that the instantiation of F by O cannot constitute a ground for F's continuous existence without some further property of O also constituting a causal basis for F's manifesting.

In support of this, consider that when F is not manifesting, it is natural to say that O is disposed to manifest in such-and-such a way. But what makes it the case that O is so disposed? Here is one answer: O simply has F. But what does it mean to say that O *has* F, if not to say O has the property of being the way of having F, and having the property of being this way, W, is a categorical property. But given the tight connection between W and F, the worry is that W is causally relevant to F's manifesting. So F is not a pure disposition. A second answer: O is the kind of thing that has F. So, O has the categorical property of being an instance of the kind of thing that has F. This seems to repeat the view that F is a way that O is and this takes us back to the response to the first answer. A third answer: O has F in virtue of the instantiation relation between O and F. But O having this instantiation relation is a categorical property of O, and threatens to serve as a causal basis for F. For what is the nature of the instantiation relation? Suppose the instantiation relation is supervenience, then what is the supervenience base? It must be some intrinsic property of O, and this implies that the purity of F is lost. Because supervenience is a covariance relation, it seems that the supervenience base will form a causal basis for F.[19] Thus, by constantly having to invoke some non-F property of O to account for O being the way it is, the non-F property becomes not just a ground for F's being but also part of the causal basis of F, since the non-F property will be causally relevant to F's manifestation.

---

[18] On Lowe's Four-Category Ontology (2006), dispositions are not modes or tropes, but universals, and universals characterize *kinds* of objects. So, Lowe does not accept pure disposition tokens and so I think he would deny the Pure Dispositions Thesis. I am just using Lowe's view as a model of how (Object) might be understood.

[19] Suppose the instantiation relation is identity (that is, F is its own instantiating property in O). Then F is pure, but F's being instantiated by an object makes no ontological difference—F grounds itself, exactly the theory to be elaborated below.

If F is grounded in its object-bearer by just being a way O is, then when F is not manifesting there remains a question about how O instantiates F, or what it is about O such that it has F, or what F is doing when not manifesting. This issue haunts all of the possible answers mentioned in the previous paragraph. The underlying idea behind the critique being advanced against (Object) is that if properties are ways objects are, then this way will be inextricably tied to this object (if not, then F could float free, and have no causal basis, and thus an object is not the grounds for F). This object will have the property of continuously existing, thus grounding F's being too. But whatever makes O the way that it is, if it is some further property beyond F, as it plausibly must be, then that property will be caught up in the causal process of F being stimulated. Thus, F will have a causal basis and not be pure.

One might retort that on (Object), it is simply a brute fact about O that it has F. If so—if this way O is does not subtly assume some further property that threatens to be a causal basis for F—then the pressing concern seems to be how O *grounds* F, the purpose of (Object) in the first place. If F is grounded by O, and grounding is a relation between properties, then some further property must be involved in the grounding base; then the worry is that this further property is causally relevant to F's manifesting, i.e., that it constitutes a causal basis for F.

This is problematic on *any* non-self-grounding account of pure dispositions, not just (Object), although it seems especially worrying on (Object) because the further grounding property is intrinsic to the very object that possesses F, unlike on other theories discussed above that invoke extrinsic grounding properties. It is true that I argued in Section 2 that grounds and causal bases can come apart in intrinsic cases as well as extrinsic cases. But the kind of case I envisioned there was a case where each property of a bundle object was grounded in the other properties of the bundle, and where each property had a distinct existence apart from the bundle. But in evaluating (Object), I have explicitly assumed a non-bundle theory of objects where the properties of objects are ways of the object, and not separable from them.

In sum, I tentatively conclude that (Object) is not a satisfactory account of the being of pure dispositions. If properties of objects are *ways* objects are, then if F is stimulated, so must be some further property of O; thus F has a causal basis. Note that my critique of (Object) applies straightforwardly to (Blobject) as well, to the extent that the Blobject is a substantial object and not a bundle object.

I want to return briefly to the bundle theory of objects as a possible account of F's being. The idea is that elementary particles are just bundles of compresent pure disposition tropes; e.g., electrons on this view would just be bundles of mass, charge, and spin. The other dispositions of a bundle-object ground a given F; e.g., mass-charge grounds spin, spin-charge grounds mass, etc. Do these other dispositions form a causal basis for F's manifestation, thus violating Criterion 2? I argued in Section 2 that this is not the case, in arguing that grounds and causal basis may come apart (Assumption 2), using the bundle theory as a possible case of this (a *mass-charge-spin* bundle object). But even if I am right about this, I am reluctant to invoke the bundle theory because of other problems. First, there is the possibility of having 'free' mass, charge, or spin; i.e., these properties may be able to exist alone or float free, apart from the bundle objects that bear them. Second, if there are other options to account for the being of F that do not assume a particular view of objects, then it is dialectically advisable to pursue those options.

Other accounts have also rejected objects as grounds for the being of pure dispositions. Molnar (2003: 151-2) thinks that the properties of elementary particles are pure dispositions, but he holds that the objects that bear these pure dispositions are point-size elementary particles. So there is no *object*, properly speaking, if objects require extension. Mumford (2006) thinks that since fundamental particles have no parts (they are simple), and since fundamental particles have no other properties beyond their dispositions, this means that the fundamental dispositions do not supervene on either parts or properties of elementary particles. It seems that the elementary particles are just bundles of dispositions on this view. But whatever the object is, it is not something that grounds the being of pure dispositions.[20]

The upshot is that pure dispositions seem to be ontologically independent of objects; if they are instantiated by objects it is a contingent instantiation. I have argued that (Global), (World), (Object), and (Blobject) as a sub-theory of (World) and (Object), do not satisfactorily solve the Problem of Being. The theory that I think best answers the Problem of Being maintains that pure dispositions are self-grounded.

6   The Self-grounding Theory

The theories evaluated thus far appear to satisfy Criterion 1, while Criterion 2 and other considerations put pressure on the theories. If we interpret the Problem of Being as a question of what a pure disposition, F, is doing when it is not manifesting, then it is plausible that none of the theories satisfy Criterion 1. Moreover, it seems that coming to grips with the Problem of Being consists in explaining how F continues to exist when un-manifested *regardless* of what is happening outside the spatiotemporal boundaries of F itself and regardless of the status of other entities. One might think this is so because it seems there could be a world in which there is only one entity, a solo pure disposition; this is consistent with the self-grounding theory, but not the other theories which

---

[20]Williams (2009), arguing against Mumford (2006), suggests that particular elementary particles must *be some way* at all times, and this is nothing less than having a categorical property. If this way an object is, is categorical and grounds a 'pure disposition', then that pure disposition would not be pure.

look beyond F itself.[21] If a solution can be given to the Problem of Being that satisfies Criteria 1 and 2, without the problems of the other theories, then that account holds favor.

As the Problem of Being indicates, for any pure disposition, F, a satisfactory theory should explain F's continued existence (i.e., the ontological grounding of F) without invoking additional categorical or dispositional properties that constitute a distinct causal basis for F's manifestation, lest F not retain its purity.

Crucial to the self-grounding theory is the assumption that F can be its own causal basis, as defended by McKitrick (2003a) and introduced as Assumption 1 (in Section 2).[22] When a manifestation occurs, it is F itself (and not some additional property that F supervenes on, or is realized by, etc.) that is stimulated, triggering F to manifest. Hence, F does not have a *distinct* causal basis. This is important to the self-grounding theory since some explanation is needed concerning what is causally relevant to F's manifesting.

So how is it that a pure disposition, F, is ontologically grounded, but nonetheless F does not have a distinct causal basis? The correct answer, I propose, is that (i) F is its own causal basis (so the causal basis is not distinct from F) and (ii) F is its own grounds for being. So the self-grounding theory is simply:

> (Self-grounded) Any pure disposition, F, grounds itself and is thus
> solely responsible for its continuous existence.

The theory of self-grounding holds that F accounts for its own continuous existence, without invoking any properties that would constitute a distinct causal basis for F; so, F remains pure as required by the Pure Dispositions Thesis. Mumford (2006: 485) similarly answers the Problem of Being, arguing that a pure disposition is grounded in 'Nothing other than itself. It grounds its own manifestations.' However, in giving this answer he affirms McKitrick's idea (that F can be its own causal basis) and then seems to assume that this implies that F is self-grounded, without giving any further explanation of the phenomenon of self-grounding as it pertains to dispositions. Similarly, Handfield (2008: 306) suggests that a pure disposition is identical to its causal basis in the context of trying to account for its grounds for being.[23]

However, if F is its own causal basis that does not necessarily mean that F is its own grounds for continuously existing. This is because the causal basis of a disposition and the grounds of a disposition may be different, as argued in Section 2. Importantly, this may be true of pure dispositions too: e.g., al-

---

[21] However, a solo pure disposition is consistent with the priority monistic version of F's grounding too. But, in that case the World = F, so F is simply self-grounded.

[22] So, a pure disposition (or 'bare disposition' in McKitrick's terminology) has no causal basis either in further dispositions or categorical properties, but it is its own causal basis.

[23] Handfield (2008) does not necessarily affirm that this is the best solution to what I call the Problem of Being. In fact, he proposes two competing theories: what I call (Global) and (Self-grounding) in Sections 3 and 6 respectively.

though a pure disposition might be its own causal basis, it may be ontologically grounded extrinsically (where the grounds are not part of the causal basis), or in other ways evaluated in Sections 3, 4, and 5. Thus, F's self-grounding does not necessarily follow from the fact that F is its own causal basis, as Mumford and Handfield suggest. To convincingly argue for the self-grounding theory—thus yielding a triple identification between the causal basis of F, the grounds of F, and F itself—one needs to rule out other grounding options. Thus, supposing the evaluation conducted in previous sections has effectively ruled out other grounding options, we can tentatively infer that (Self-grounded) is correct. Additionally, considerations of ontological simplicity seem to favor the self-grounding of F, at least as compared to options that invoke an additional ontological category of substantive objects (understood as substances instantiating properties). Regardless, whether or not I have effectively ruled out other grounding options, (Self-grounded) remains a legitimate contender worthy of examination.

Suppose that F is self-grounded. It is not clear how this accounts for F's continuous existence. That is, *how* is F self-grounded? How does F account for its own continuous existence through periods of non-manifestation? I will next develop a principle concerning how pure dispositions are self-grounded. Though having a viable explanation of F's self-grounding does not completely rule out other grounding options, it at least enhances the plausibility of (Self-grounded).

7   The Principle of Minimally Sufficient Occurrence

How does F ground itself? What does it mean for F to be self-grounded? I suggest the following principle in support of (Self-grounded):

> Minimally Sufficient Occurrence: A pure disposition, F, is self-grounded if and only if F undergoes a minimally sufficient occurrence of F's own power when F is not engaged in one of its more characteristic manifestations.

This principle is a supposition to explain F's continuous existence or being when F is not manifesting one of its other possible, and generally characteristic, manifestations. The characteristic manifestations are those manifestations comprising F's causal role in a system of dispositions (more on this below).

One motivation for this principle is the idea that we should look to the nature of F itself for an explanation of F's continuous existence, already implicit in the idea of self-grounding. In looking to the nature of F, we are drawn to the power-hood of F. The core idea is then that F's persistence or continuous existence lies in manifesting a minimally sufficient range of its power.[24] By 'range'

---

[24] On this view, persistence is dispositional, not categorical; cf. Williams's (2005) concept of static dispositions.

I mean the total set of possible manifestations F can undergo. So, this proposal assumes the following:

> Assumption 3: Any token disposition, F, is multi-track, such that F can manifest in multiple ways depending on the manifestation circumstances.[25]

This means that a token disposition may receive multiple kinds of stimuli and thus may manifest in multiple ways. In other words, dispositions possess *the power* to manifest in many ways. As such, an instance of a pure disposition has a set of many powers (Mumford 2004: 171). Assuming pure dispositions are multi-track, a minimally sufficient occurrence is just *one* of the possible ways F can manifest, so F's total dispositional nature is much more than its capacity for minimal manifestation. Thus, while F does not nearly manifest all it is capable of at any given time, F does manifest some of its power thereby continuously existing and staying ready for future characteristic manifestations. Mumford (2006: 485) raises the Problem of Being when he asks: 'in what, actual, does an unmanifested, elementary casual power [i.e., pure disposition] consist?' Similarly, Psillos (2006b) asks what pure dispositions do when they are not manifesting. The answer I am proposing is: Pure dispositions are not ever in a non-manifesting state—yet they *are* dispositions. Thus, my proposal rejects the condition assumed in the Problem of Being that any pure disposition may be in a completely latent state.[26]

I will now develop my proposal by answering five questions about it, including whether a minimally sufficient occurrence implies that pure dispositions are categorical properties.

The first question is: *What kind of manifestation is a minimally sufficient occurrence of F?* Perhaps it is best to cast the answer negatively: The minimally sufficient occurrence of F is *not* a characteristic manifestation of F. The characteristic manifestations of F are those possible manifestations related to the causal role F typically occupies in a system of dispositions that includes some of F's disposition partners, i.e., dispositions that may trigger F and that F may trigger (Heil 2003: 11). This can be differently understood from an epistemological perspective: the characteristic manifestations of F are typically those by which we, observationally or theoretically, identify and define F, based on the variety of possible stimuli that in fact, or might, trigger F. But we don't identify or define F by its minimally sufficient occurrence, so it is not a characteristic manifestation in the sense given. Suppose mass is a pure disposition;

---

[25] Ryle (1949) introduced multi-track dispositions, and others have posited them, e.g., Martin (2008) and Mumford (2004). Such dispositions are capable of manifesting in different ways given the variety of stimuli they are subjected to; e.g., a token of fragility might manifest as shattering, cracking, or chipping. The minimally sufficient occurrence of a pure disposition is just one track of its power.

[26] Maybe the proposed theory of self-grounding only answers a narrow interpretation of the Problem of Being, concerning what F is doing when not manifesting. In that case, the proposal at least presents a narrowly defined solution worthy of examination.

mass is identified and defined in terms of its causal role with other instances of mass and other relevant dispositions, such that these resultant manifestations, but *not* a minimally sufficient occurrence of mass, constitute its characteristic manifestations.[27]

However, this account of F's characteristic manifestations does not imply that the minimally sufficient occurrence of F is not subsumed by the complete *causal profile* of F, for a minimally sufficient occurrence does have causal significance—specifically, in grounding F. Two points, however, differentiate the causal roles related to all of the characteristic manifestations of F, and the causal role F has in maintaining its being via a minimally sufficient occurrence. First, the latter is not the kind of causal role typically used in identifying and defining F since *all* pure dispositions, if this theory is correct, will have such a power to occur minimally sufficiently to ground their being, and second, F's minimally sufficient occurrence is not the result of the interaction of F with a disposition partner as F's characteristic manifestations are.[28]

Given these details, here is a more precise statement of the relation between the minimally sufficient occurrence of F and F's characteristic manifestations: Given that F is multi-track, and has the power for both characteristic manifestations and a minimally sufficient occurrence, then F manifests its minimally sufficient occurrence during any segment of time, T, so long as during that segment of time, F is not activated by a stimulus, S, that triggers F to undergo a characteristic manifestation; and if S does trigger F into a characteristic manifestation, then when the characteristic manifestation stops, F reverts to its minimally sufficient occurrent state. So, F need not minimally sufficiently occur at all times, but only when F is not manifesting characteristically. Indeed, F need not *ever* manifest in a minimally sufficient way, supposing F is constantly involved in characteristic manifestations, as may be true of the dispositions of fundamental particles, for example (often cited by proponents as examples of pure dispositions).

So, a minimally sufficient occurrence is not an occurrence or manifestation of a kind that is characteristic for F, but it is consistent with the kind of disposition that F is, for it falls within the total cluster of possible manifestations of F. If F is mass, for instance, then the minimally sufficient occurrence necessary for mass' being is, indeed, a manifestation of mass. F's minimally sufficient occurrence means that F is undergoing the process of manifesting some of what it can do and it is this fact that accounts for F's continual existence when not engaged in its characteristic manifestations.

---

[27] Given the distinction between 'characteristic manifestations' and the 'minimally sufficient occurrence' or manifestation, one might just call the latter 'non-characteristic manifestation'; however, the former phrase captures the idea that the kind of manifestation picked out is *sufficient* for the continuous existence of F.

[28] A minimally sufficient occurrence of F *may* be insufficient for detecting F, unlike F's characteristic manifestations. That is, our best possible observational techniques may indicate no manifestation of F even when F is undergoing a minimally sufficient occurrence.

As an analogy, suppose the familiar disposition fragility is a pure disposition. A characteristic manifestation of fragility would be cracking, for instance. But a minimal manifestation of fragility would be a case of the ever-so-slightest cracking or a prolonged (over days or longer) cracking, perhaps undetectable to unaided human observers. The suggestion is that fragility continuously manifests itself thus maintaining its being, yet it is continually capable of manifesting in many more ways.[29] It is important to note that because fragility has a distinct causal basis, the minimally sufficient occurrence of fragility requires that its causal basis be triggered by a stimulus typically associated with fragility's characteristic manifestations. This is unlike pure dispositions, on the theory being offered. This is because—assuming characteristic manifestations are generally associated with what we might call characteristic stimuli—the minimally sufficient occurrence of F does not require a characteristic stimulus.

With the core of the theory now in place, one might object that any property X needs to exist (be instantiated) in order to do something, whereas my claim implies that X's doing something gives X existence; so F's existing in virtue of its minimally sufficient occurrence (a type of functioning) seems to incorrectly reverse the order of existence and functioning. Temporally, existence comes before functioning, whereas my account implies that functioning comes first. However, this is not what my theory claims. Rather, my theory is that F's continuous existence just *consists in* its functioning (i.e., displaying its power): existence and functioning are packaged together in pure dispositions. The series of manifestations F undergoes, including its minimally sufficient occurrences, is a dynamic process of continual dispositional self-generation.

The second question is: *If the minimally sufficient occurrence is an event (a manifestation event), how does an event ground a property?* In other words, given that properties generally ground properties, how is it that an event (i.e., a minimally sufficient occurrence) accounts for the grounding of F?

The manifestation event does *not* ground F; rather, the grounding of F's being just *is* a manifestation event. To explain this answer, following Kim (1998) I will suppose that events are property-exemplifications: an event, E, $=_{df}(x, P, t)$, an object x exemplifying a property P at a time slice t or temporal interval $t_1 \ldots t_n$. (If F is not borne by an object, then x drops out of the definition, or perhaps spacetime exemplifies F.) Assuming that this theory of events is right, F's manifestations are property-exemplifications. Thus, the event of F exemplifying one track of its power at any given time t or from $t_1 \ldots t_n$, i.e., undergoing its minimally sufficient occurrence, is the basis of F's being or continuous existence over time.

---

[29]It is not so implausible that fragility is continuously manifesting some of its power, if one considers that a fragile glass is constantly bombarded with particles, dirt, etc. So despite appearances it is not obviously false that the glass is minimally manifesting its fragility by slowly breaking over an extended time. Of course, this does not mean that fragility needs to manifest minimally in this way, whereas on my theory pure dispositions do need to manifest minimally sufficiently to continuously exist.

On the proposed solution to the Problem of Being, F's self-grounding via a minimally sufficient occurrence is an event that keeps F remaining ready for further exemplifications of its power. If a given minimally sufficient occurrence of F is temporally extended, then it is a temporally extended event, or process—i.e., a powerful or dispositional process. An interesting implication of this event-theory of F's self-grounding, is that if pure dispositions are fundamental properties then the fundamental entities or constituents of the world are events.

The third question is: *What happens when F manifests in a characteristic way?* Then, the minimally sufficient occurrence stops.[30] Thus, there are (nearby) possible worlds in which F is not manifesting a minimally sufficient range of its power, i.e., worlds in which F manifests in one of its characteristic ways. Thus, F is not continuously manifesting *all* it is capable of in all possible worlds (if it were, this would seem to make it a categorical property). When F displays one of its characteristic manifestations, then it is not minimally sufficiently occurring. Thus, the fact that F undergoes a minimally sufficient occurrence when not undergoing a characteristic manifestation does not make F categorical because it need not always undergo a minimally sufficient occurrence, just as the other possible manifestations of F are occurrences of F and do not make F categorical.[31]

Still, one might press that because F is continually manifesting *some* of its disposition—or something of which it is capable—this makes it categorical. But why is this so? A categorical property, strictly speaking, is not capable of manifestations as dispositions are. A categorical property is what it is at any given time, whereas a disposition is full of possibilities, even while manifesting some power (although there are some dispositions like fragility that release all of their power, so to speak, on some of their manifestations).

The fourth question is: *What is the activating condition (or stimulus) for F's minimally sufficient occurrence?* Assuming that characteristic manifestations of F are generally associated with characteristic stimuli, the minimally sufficient occurrence of F does not require a characteristic stimulus.

However, it may not even require a non-characteristic stimulus, for perhaps F's minimally sufficient occurrence is spontaneous or self-generated (akin to radioactive decay). On that view, then F is its own stimulus for its minimally sufficient occurrence. But if a stimulus is required, perhaps it is constituted by negative conditions: F being in the absence of stimuli appropriate for F's characteristic manifestations. Yet another possibility is that F being

[30]We might call the minimally sufficient manifestation state of a pure disposition its 'static-side' as opposed to its more 'dynamic-side', to differently employ the distinction between static and dynamic dispositions introduced by Williams (2005). (He employs the distinction for instances of different kinds of disposition, whereas I am suggesting it be used for different tracks of power of a token disposition.)

[31]Some pure dispositions may be gone forever once they undergo a certain type of manifestation, while others may retract back to a minimally sufficient occurrence state after any of their other possible manifestations occur.

situated in spacetime stimulates it to undergo its minimally sufficient occurrence (implying that F could not exist sans spacetime, which seems true for all concrete entities). For example, if mass is a pure disposition then perhaps its minimally sufficient occurrence is the bending of spacetime.[32]

The fifth question is: *Does a minimally sufficient occurrence of F make F categorical?* This is important because it questions the power-hood of pure dispositions. On the theory I am defending, pure dispositions continually manifest their power. In contrast to Psillos (2006b: 141), I am claiming that continually manifesting dispositions may still be dispositions, not categorical properties. Once we accept that pure dispositions need not manifest everything they are capable of at any given time, we can allow that they undergo a minimally sufficient occurrence during stretches of time when they are not manifesting in more characteristic ways.

However, suppose that F undergoing minimal occurrence indeed implies that F is not a disposition but a categorical property. Then, if the proposed theory is the best available response to the Problem of Being, then the overall argument of this essay should be construed as a *reductio* of the Pure Dispositions Thesis. This would be a significant conclusion in its own right. On the other hand, if a minimally sufficient occurrence of a pure disposition does not imply that pure dispositions are categorical properties, then we have a viable theory of the being of pure dispositions.

I maintain that we can avoid the problem—that F's continuous manifestation of any sort along F's possible lines of manifestations implies that F is categorical—by attending to the distinction between 'occurrent' and 'categorical'. Sometimes these terms are used interchangeably. However, by an 'occurrent' property I mean a disposition in a state of manifesting—a disposition is *occurring* (or, manifesting). This is a dynamic process. As with all manifestations F may undergo, the minimally sufficient occurrence of F does not necessarily make it categorical. A categorical property like having shape is supposed to be static, possessing a quality of 'just-there-ness' (Armstrong 2004: 141). A categorical property may be a leftover of some causal process involving dispositions in the past—a glass' microstructure, for example. But an *occurrence* of a pure disposition, or any disposition, is the process of the disposition actively manifesting.

Categorical properties are complete in the sense that they fully 'manifest' all they are capable of at any given time. By contrast, dispositional properties are not fully manifest. Each manifestation of a disposition, including a pure disposition, just taps the surface of potentiality built into it. This is why they are modal properties. Hence, F continuously manifesting some of its power (whether a minimally sufficient occurrence or a characteristic manifestation) does not imply that F is categorical. F's total multi-track power remains intact even while manifesting some of its power. Thus, since pure dispositions do not display all they are capable of at any given time, to say they are categorical like

---

[32]Thanks to Luke Elwonger for this example.

traditional categorical properties is misleading. F is not categorical because it is always full of threats: even as it is manifesting in one way, it is capable of manifesting in another way. At the most, some limited range of F's power is displayed at any given time, which is to say that the disposition is undergoing a process of manifesting, not that it is categorical.[33]

The opponent of the Pure Dispositions Thesis might respond that a minimally sufficient occurrence of any pure disposition implies that it has a categorical aspect, as according to the dual-aspect theory found in Martin (2008), or that it implies that any pure disposition is identical to a categorical property (Heil 2003); if either view is true, then the Pure Dispositions Thesis is false. It is worth noting that on both of these views properties have dispositional natures and are identified in terms of their causal roles. A prime motivation for adding a categorical dimension, to what are otherwise fully dispositional properties, is to account for their continuous existence (the idea is that to say a property X is 'categorical' is to say that X is always there). Despite this, restricting discussion to the fundamental properties on the dual-aspect view or identity thesis, it is plausible that it is the dispositional nature of these properties, *not* the added categorical dimension, in virtue of which they continually exist. This is because continuous existence appears to be a causal process.

Rather than invoking a categorical aspect or an identical categorical property to explain the continuous existence of what would otherwise be pure dispositions, my account offers an explanation of the being of pure dispositions that invokes the powerful nature of such properties. On my view, pure dispositions are purely powerful and continuously manifesting some of that power. Perhaps this is similar to the sense of categoricalness sought by the dual-aspect or identity theorist. But my account differs from these views in that F's being is explained by reference to F's power, i.e., by reference to F's dispositionality.

8   Conclusion

In sum, this paper proposes that pure dispositions are self-grounded in virtue of continually manifesting a minimally sufficient range of their total multi-track power. Thus, pure dispositions continuously exist because of their own power and thereby remain ready for all their other possible manifestations.[34]

---

[33] Still, one might maintain along with Mumford (2006: 485), who accepts the Pure Dispositions Thesis, that pure dispositions are just as 'categorical' as traditional categorical properties. That is, pure dispositions have a categorical existence just as categorical properties do, yet throughout their existence they are potent.

[34] Acknowledgements: I would like to thank Jennifer McKitrick for many helpful comments and fruitful discussions about the ideas in this paper. I would also like to thank participants at the Graduate Student Colloquium at the University of Nebraska–Lincoln on 2 October 2009. Finally, I would like to thank the audience at my presentation at the Central States Philosophical Association meeting on 9 October 2009, for helpful comments and questions, and in particular Ronald Loeffler for his very useful commentary on my paper.

Part V

PAN-DISPOSITIONALISM

THE METAPHYSICS OF PAN-DISPOSITIONALISM

Matthew Tugby

1   Pan-dispositionalism

In this paper I will focus on the metaphysical view about natural properties known as pan-dispositionalism. My aim is to address some important metaphysical questions about the pan-dispositionalist picture, so that we may begin to see how pan-dispositionalism is best understood. Before introducing these questions I must briefly outline the central claims of pan-dispositionalism.

During recent years, several philosophers (for example, Ellis 2001, Martin 1993, and Molnar 2003) have advocated a metaphysical view which sees at least some natural, sparse properties as being irreducibly dispositional (or 'powerful') in nature. According to this view, the natures of several (if not all) properties are determined, at least in part, by the causal abilities that they bestow upon their possessors. Such a view sets itself against the previously dominant categoricalist views according to which properties are essentially inert qualities whose natures are independent of facts about causality. According to categoricalist views, such as those of Armstrong (1997) and Lewis (2009), properties bestow causal roles merely contingently.

Now, it is a striking feature of dispositions that their natures are determined by what they are dispositions *for*, i.e., by the manifestation property towards which they are directed. For example, to understand the nature of charge, which is a paradigmatically dispositional property, one must know how charged particles behave in certain circumstances (e.g., they accelerate when placed in a force-field). Pan-dispositionalism is the strongest form of realism about dispositions, for on this view *all* natural properties (and relations) are irreducibly dispositional in nature. The pan-dispositionalist is thereby committed to a holistic metaphysics in which all properties 'form an interconnected web' Mumford (2004: 182). It is holistic in the sense that the nature of a given property will be determined, at least in part, by its directedness towards a further property, but given that this further property will be irreducibly dispositional in nature, its identity will also be fixed by the manifestation property towards which it is directed (and so on).

Before introducing the questions that I want to address with respect to pan-dispositionalism, it should be pointed out that pan-dispositionalism comes in slightly different forms. This is a point that is often not duly acknowledged. Pan-dispositionalism is often closely associated with dispositional monism, but these two views are not equivalent. According to dispositional monism, the nature of each and every property is *exhausted* by its dispositional characteristics (see, for example, Bird 2007a and Mumford 2004). Whilst dispositional monism clearly entails pan-dispositionalism, pan-dispositionalism does not entail dispositional monism. One could consistently maintain that each and every property has an irreducible dispositional nature, and thereby sign up to pan-dispositionalism, without claiming that *all* there is to a property is its dispositional characteristics. Such a position seems to have been occupied by C. B. Martin (1993), when putting forward his 'two-sided' view. According to the two-sided view, all properties have a categorical (or 'qualitative') aspect to them, but at the same time they also have an irreducible dispositional nature. Such a view is therefore pan-dispositionalist in spirit, and so falls under the scope of this paper, along with dispositional monism.

2   The aims of this paper

During this paper I will assume the pan-dispositionalist picture and begin by addressing the question whether properties are best understood as universals or (sets of) tropes. After briefly introducing the distinction between tropes and universals, I will approach the tropes versus universals debate by considering whether and how each of these views about properties are able to accommodate and explain certain salient features of irreducible dispositionality. More precisely, I will claim that any satisfactory version of pan-dispositionalism must provide an account of the *directedness* of dispositions, whilst accommodating the fact that many (if not all) dispositions are *intrinsic* to their possessors and also the related fact that a disposition instance may exist unmanifested.

I will begin by considering how the 'universals' version of pan-dispositionalism is able to accommodate these important facts, before examining 'trope' versions of pan-dispositionalism. My conclusion will be that a theory of universals is able to provide a more coherent and transparent account of these central features of irreducibly dispositional properties. This does not, of course, mean that the 'universals' pan-dispositionalist is entitled to declare an immediate victory, since as is well known, the tropes versus universals debate may be contested on a variety of grounds. What I will argue, however, is that if one signs up to pan-dispositionalism, then one has special reasons for favoring a universals account, reasons that, at the very least, put the onus of proof on those seeking to establish a trope version of pan-dispositionalism. Whether these reasons can be trumped on other grounds is a question I leave open.

The 'universals' version of pan-dispositionalism to be recommended is one that sees universals as being, at least in part, *relationally constituted*. On this view, disposition universals are internally related, and it is in virtue of such relations that a disposition's directedness is what it is. Such relations are what Bird (2007b) calls second-order 'manifestation' relations. I will conclude this paper by briefly considering the nature of these relations.

Before beginning the argument, it should be pointed out that although I am framing the tropes versus universals debate in terms of pan-dispositionalism, the reasons put forward in favour of a universals account should appeal equally to those who, whilst allowing a place in their ontology for irreducible dispositionality, do not claim that *all* properties (and relations) are irreducibly dispositional. Examples of such views are found in Ellis (2001) and Molnar (2003).

Finally, it should also be noted that whilst I will be recommending a universals account of dispositionality, I will not address here the question whether disposition universals are best understood in the Aristotelian 'immanent' sense, or the Platonic 'transcendent' sense.[1] Unfortunately, that question must be reserved for another day.

3   The tropes versus universals debate

Those who hold there to be a distinct ontological category of natural properties (and relations) typically view those properties (and relations) as either universals or as (sets of) tropes. An initial way of capturing the difference between universals and tropes is to say that universals can exist in many places at the same time, whereas tropes cannot. Thus, if properties are universals, then all objects that exemplify, say, a particular determinate shade of red may be said to share an *identical* property; the very *same* determinate property is exemplified by all of those objects. On this view, a property is therefore an entity that can spread itself across many concrete particulars. In contrast, according to the trope theory, each instance of a property is *distinct*, which is to say that properties are nonrepeatables. On this view, each property instance is itself a particular, although since property instances cannot exist apart from their

---

[1] According to the immanent view, universals wholly exist in the space-time realm, in each of the particulars that instantiate them (they exist *in rebus*). In contrast, according to the transcendent view, universals (or 'Platonic forms') exist in a realm that transcends space and time (they exist *ante rem*).

possessors, they are abstract particulars. On the trope view, then, the claim that a group of distinct objects share the same determinate property should, strictly speaking, be understood as the claim that those objects each possess distinct property instances which resemble exactly.

Which view of properties should the pan-dispositionalist favor? As suggested earlier, one way of choosing between these alternatives is to consider which of these views offers the best resources for accommodating certain salient facts about irreducible dispositions. The facts about dispositions I will focus on concern the *directedness* of dispositions, the *intrinsicality* of many (if not all) dispositions, and the related fact that an instance of a disposition may *exist unmanifested*. At first glance, these facts appear to be at odds with each other, but with universals in play, one can, I argue, accommodate these facts in a coherent and transparent way. In contrast, it is less clear that a trope pan-dispositionalist can simultaneously accommodate these facts in a satisfactory way.

4   Three facts about irreducible dispositionality

Directedness.

First, as mentioned already, it is a fact about dispositions that they are in some sense connected with, or 'directed towards', that which they are dispositions for, i.e., their manifestation property. This is a key fact, for as we saw earlier, it is in virtue of such directedness that the identity of a disposition is fixed. To know the nature of charge, for example, is to know what outcomes being charged is orientated towards.

Intrinsicality.

The second important fact about dispositions is that at least some dispositions instances are intrinsic to their possessors. Defining the term 'intrinsic' in a precise way is no easy matter (see Langton and Lewis 1998). However, a rough-and-ready definition capturing our main intuition about intrinsicness is all we need for our purposes. Our main intuition seems to be that a property P is an intrinsic property of x if and only if x's having P is independent of the existence of distinct entities and x's relation to them. Do any dispositional properties satisfy this definition? It seems that they do. The negative charge possessed by, say, an electron is surely a feature that it has independently of the situation external to the electron. If a particle is negatively charged, it would remain charged even if put in very different circumstances (unless, of course, the particle itself was to be changed in some way).

It should be pointed out that some philosophers have argued that there are dispositions which are not *in*trinsic, but *ex*trinsic. McKitrick (2003a) offers the property of weight as one such example. If a person is moved from one planet to another, their weight may change, even if the person remains intrinsically identical. This suggests that the dispositional property of having a certain weight is an extrinsic one. It seems clear enough, however, that not all dispositional properties are of this kind. When we come to *explain* why a person's weight would be different if they lived on a different planet, we inevitably appeal to properties that do seem to be intrinsic. To understand weight, for example, is to understand that it is a function of the person's mass and of the magnitude of the gravitational field generated by the planet's mass. In contrast to weight, mass is plausibly an intrinsic dispositional property, because no matter where a massive object is located, it will have the same set of gravitational abilities. It seems, therefore, that intrinsic dispositions cannot be eliminated. As Molnar puts it, '[S]uch is the resilience of the intrinsic' (2003: 107).

Given that weight can be explained by the mass of a person along with the gravitational field in which they find themselves, one may suspect that having a certain weight is really no addition of being, and that such a 'property' can be explained away. This is Molnar's suspicion (2003: 108–10), but McKitrick (2003a) has a number of responses to this line of argument. This debate need not concern us here, however. The important point is that at least some dispositions are wholly intrinsic to their possessors.

Existence unmanifested.
The third fact about irreducible dispositionality to be considered, which is related to the intrinsicality fact, is that an instance of a disposition may exist even if it is never manifested. Acceptance of this fact is central to the realist view about dispositions. According to the realist view, dispositions are properties in their own right and so may exist even if they are not being displayed. Whilst the manifestation of a disposition is potential only, the disposition itself is *actual*. Given that this is so, the fragility of a particular vase, for example, would be ascribable to it even if the vase never comes to be broken.

5   The challenge of accommodating the three facts

There is, I suggest, a *prima facie* tension between these three facts. Specifically, there is tension between the first fact, about directedness, and both the second and third facts, which concern intrinsicality and existence unmanifested. This tension is revealed as soon as we consider how one might go about accounting for the directedness of irreducible dispositions.

U. T. Place (1999: 226) once remarked that when we cash out the directedness of a disposition '...we are characterizing it in terms of its 'relation' to something...'. This quote suggests an obvious way of accounting for the connection between a disposition and its manifestation. Such an account would ground the directedness of a disposition towards its manifestation in a genuine relation. This does indeed seem the obvious way to go. When we say, for example, that the thigh bone is connected to the knee bone, what we ultimately mean is that the thigh bone bears a certain relation to the knee bone.

The first problem with this idea, however, is that since the nature of ir-reducible dispositionality consists in *nothing more* than directedness, dispo-sition instances would become purely relational features of the world. This result is at odds with our second fact, that many disposition instances are in-trinsic (i.e., monadic). If we take dispositions to be purely relational entities, it seems, at first glance, that the intrinsicality fact is compromised. If, on the other hand, the relational account of directedness is rejected, so that the in-trinsicality fact may be preserved, we are left in the dark with regard to what dispositional directedness consists in.

The second problem is that the relational view of directedness also seems at odds with the third fact, that a disposition may exist unmanifested. This is a worry that Place is well aware of. It is noticeable in the quote above that Place uses inverted commas when speaking of there being a 'relation' between a disposition and that which it is a disposition for. This suggests he holds some skepticism about the idea, and the reason is that it seems to compromise the third fact, that a disposition instance may exist even though its manifestation never occurs. The problem here is that, intuitively, in order for a relation to exist, its relata must also exist. But in the case of an unmanifested disposition, one of the relata in question is missing, since the manifestation towards which the disposition is orientated never occurs. This is what some have called the Meinongian problem (see Armstrong 1997: 79).[2] If fact three is upheld, the re-lational account of directedness is in trouble, unless one is prepared to take the radical step of accepting relations which lack relata. Alternatively, one could simply reject the relational view of directedness, but then we are once again left in the dark with regard to what dispositional directedness consists in.

In sum, then, the challenge the pan-dispositionalist faces is that of pro-viding a theory about dispositions which can account for directedness in an intelligible way whilst at the same time preserving the fact that many disposi-tions may be instantiated intrinsically and may also exist unmanifested. I will now argue that if the pan-dispositionalist views properties as universals, this challenge can be met straightforwardly.

6   Universals to the rescue

With universals in play, one has the option of cashing out the directedness of a disposition in a relational way, by appealing to *second-order* manifesta-tion relations holding between the disposition universal and the universal(s) corresponding to its manifestation(s).[3] If one is a pan-dispositionalist, this re-lation will be seen to constitute, at least in part, the nature of the universals related. This means the second-order relation, which grounds directedness,

---

[2]This objection is so-called because, famously, in a debate with Russell about reference, Meinong appeared to advocate the reification of non-existent entities.

[3]This kind of view has been suggested by both Bird (2007a) and Mumford (2004).

must be internal in some sense.[4] If such a relation were not internal, but rather external, then properties would only have their dispositional characteristics contingently. This would clearly go against the tenets of pan-dispositionalism and would leave us with a view of dispositions closer to that of Armstrong.[5] Now, with these second-order internal 'manifestation' relations in play, the pan-dispositionalist is able to preserve the intuition, mentioned earlier, that when we speak of there being some connection between a disposition and its manifestation property, we mean that they are related in a certain way. But how, on this picture, can facts about intrinsicality and existence unmanifested be satisfactorily accommodated?

Once directedness is viewed in terms of relations amongst universals, the following moves become available. With respect to the second fact, that many disposition instances are had intrinsically, the universals theorist can assert that the internal relations which determine a disposition's nature exist merely at the *second-order* level of universals. This kind of relation is to be distinguished from what we may call first-order relations, which hold between particulars rather than the universals themselves. By appealing to this distinction within the theory of universals, the pan-dispositionalist can allow that a particular may instantiate a property intrinsically, even though the property *type* or *kind* is relationally constituted at the second-order level of universals.[6] In other words, although disposition instances at the first-order level may be said to be intrinsic to their possessors, the connection between a disposition and its manifestation property is nevertheless maintained due to the relations holding between the universals themselves.

What about the third fact concerning the existence of unmanifested dispositions? Again, using distinctions available within the theory of universals, one can accommodate this third fact without losing the connectedness that a disposition has with its manifestation. Although the manifestation of a disposition *instance* possessed by a particular often does not come into existence, the relevant manifestation type can still exist. This allows us to say that dispositional directedness does not consist in relations to *particular* manifestations which may not exist. Rather, directedness is secured by relations to manifestation *kinds*. Ellis was arguably the first to emphasize the importance of this idea.[7] It gets around the Meinongian problem because, if like Ellis, one holds an immanent theory of universals, the generic kind towards which a dispositional property is directed will automatically exist 'if something, somewhere, at some time, has an effect of this generic kind' (Ellis 2001: 133). And if one

---

[4]The sense in which such a relation may be said to be internal will be addressed in detail towards the end of this paper.

[5]Armstrong (1997: 220–62) calls the contingent second-order relations that bestow dispositional characteristics relations of nomological necessity, or 'N' relations.

[6]Bird (2007a: 141) also emphasises the importance of clearly distinguishing the second-order level of property universals with the first-order level of particulars.

[7]Mumford has also explored this way of responding to the Meinongian objection (Mumford 2004: 11.7).

holds a transcendent view of universals, then the existence of the property kind towards which a dispositional property is directed is automatically guaranteed, because transcendent universals are plausibly necessary existents (see Bird 2007a: 3.2.2). On either of these alternatives, then, it is easy to avoid the conclusion that dispositional directedness involves relations to entities which may not exist. This is made possible by the idea that dispositional directedness is secured by relations to the relevant manifestation universals at the second-order level.

Now that I have outlined how the 'universals' pan-dispositionalists can relieve the *prima facie* tension that arises with respect to the three facts, without giving up on the obvious way of explaining the directedness of dispositions (i.e., in terms of genuine relations), we will now see whether the trope pan-dispositionalists can do the same. I will argue that they cannot.

7   Heil's trope account

In contrast to pan-dispositionalists who adhere to a universals view, it is difficult for the trope theorists to swallow the claim that dispositions are (at least in part) relationally constituted. This is because, unlike the universals theorists, the trope theorists are unable to draw the distinction between property universals and property instances; for the trope theorists there are only distinct property *instances*. Therefore, on the trope view, the relational constitution claim amounts to the claim that all disposition instances are relationally constituted, and this seems to leave no room for ascribing dispositions to particulars which are purely intrinsic.

Heil, as a pan-dispositionalist trope theorist, rightly sees that the rejection of the intrinsicality fact concerning dispositions is unappealing. Also aware of the danger of Meinong-type objections, Heil suggests that '[T]he existence of a disposition (trope) does not in any way depend on the disposition's standing in a relation to its actual or possible manifestations …' (2003: 83; words in parentheses added for clarity). Later, Heil makes the same point in terms of truth-making: '[T]he truth-maker for 'this key would open a lock of kind K' is not the key, possible lock of kind K, and a relation between the key and K' (2003: 124). Rather, according to Heil, the powers are 'built in' to the intrinsic properties themselves: '[I]f the key 'points beyond' itself to locks of a particular sort, it does so in virtue of its intrinsic features' (2003: 124).

Given Heil's acceptance of the intrinsicality of dispositions, one might immediately think that he compromises the directedness thesis concerning dispositions. But as is clear in the last quotation, this is something that Heil certainly wants to avoid, for he rightly sees that the directedness thesis is at the heart of pan-dispositionalism. This is made clear, for example, when Heil claims that a powerful thing (a key, in this instance) 'points beyond' itself to its manifestation, and is 'ready to go' (2003: 124). The problem is, however, that if directedness is not cashed out in terms of relations, it becomes unclear as to what account can then be given of dispositional directedness. What, precisely, are the ontological grounds of dispositional directedness on Heil's view? This question becomes especially pressing when we recall the fact that dispositions may exist unmanifested. What, precisely, does it mean to say that an object with an unmanifested disposition trope 'points beyond' itself? Without further elucidation, the 'pointing beyond' claim seems to be a mere vague metaphor.

Heil does not appear to give any further account of dispositional directedness or of how such directedness is possible. Yet, such an account is I think needed, to make it clear what it means for a particular trope to be directed towards one manifestation rather than another. Saying merely that dispositional directedness is 'built in' to the intrinsic properties themselves does not shed much light on the matter, for the question at hand is how, precisely, an intrinsic physical property could indeed 'point beyond' itself to something that may not exist.

It is at this point that the universals account of dispositions is seen to have a distinct advantage over Heil's trope view, for it is able to give the 'pointing beyond' claim ontological backing in a way that Heil's does not. This difficulty is also not peculiar to Heil's version of the trope view. Martin, another trope pan-dispositionalist, faces similar worries. Like Heil, Martin maintains that dispositions are not relational, on the grounds that '[T]he readiness of something's disposition for all of this (manifestation) may fully exist although its disposition partners and mutual manifestations do not' (2008: 6; words in parentheses added for clarity). If one is a trope theorist, this does indeed seem like the sensible conclusion to draw, but then what metaphysical account can be given of the directedness of dispositions? Again, this is not a question that Martin seriously addresses. Martin simply tries to explain dispositional directedness using an array of metaphors and gestures: he speaks of 'dispositional readiness' (2008: 23), the 'would-have-been-if' of dispositions (2008: 2), the 'what for' of dispositionality (2008: 4), the 'ready to go' of dispositions (2008: 2), and dispositional 'selectiveness' (2008: 7). The result is that the precise nature of dispositional directedness is not made clear.

8   Molnar's intentionality view

Unlike Martin and Heil, Molnar is one trope theorist who does take more seriously the need to provide an account of the metaphysical source of dispositional directedness.[8] Molnar accepts that disposition tropes are not relational, and, following U.T. Place (1996), uses the notion of intentionality to try to cash

---

[8]Note that, unlike Heil and Martin, Molnar is not strictly speaking a pan-dispositionalist. He comes pretty close, however, because he takes all but spatial properties to be irreducibly powerful (see Molnar 2003).

out the nature of the directedness of disposition tropes (Molnar, 2003: 61).[9]
Molnar identifies four main features that most contemporary philosophers
take mental intentionality to have, and then argues that each of these features
are ones that dispositional states, or 'powers', also have (2003: 63–6). Briefly,
the four features are: i) internal reference to, or 'directedness' towards, an (intentional) object; ii) the intentional object may not exist; iii) the intentional
object may be indeterminate in some respects; iv) the truth of a description of
an intentional state need not be preserved under substitution of co-referring
expressions.

It should be pointed out that several arguments have been provided in recent philosophical literature attempting to show that Molnar's account fails on
the grounds that dispositionality differs in several crucial respects from mental
intentionality.[10] I need not rehearse those arguments here, however. Even if
Molnar is correct, and dispositional directedness does share the main features
of mental intentionality, it remains far from clear that Molnar's account of dispositional directedness is a significant improvement on that offered by Heil or
Martin. To begin with, Molnar's aim is not to provide a metaphysical account
of intentionality, but instead to merely point out that dispositional directedness is the same kind of directedness as that found in the mental intentional
case, whatever that may be. This becomes clear when Molnar indicates that
the concept of intentionality is simply taken to be primitive (2003: 81).

In Molnar's defense, one might think that he does go at least some way towards alleviating the apparent tension between dispositional directedness and
the fact that, for example, disposition tropes may exist unmanifested. Perhaps
Molnar's point is simply that we do not have any qualms about accepting directedness towards non-existent objects in the mental intentional case, and
so why should we feel uneasy in the physical case? Even if, as seems to be the
case, there are problems surrounding mental intentional directedness, these
are problems we already have in philosophy and so viewing physical dispositional states as intentional states does not bring any new problems to the
philosophical table.

In response, however, it should be said that this kind of move amounts to
ignoring the problems at hand rather than tackling them. Molnar's intentionality claim by itself does not, for example, settle the important issue of whether
dispositional directedness consists in a genuine relation. In one place, however, Molnar does go on to indicate that he thinks intentional directedness

---

[9] See also Martin and Pfeifer (1986), who suggest that intentionality as traditionally conceived
is not peculiar to the mental.

[10] Bird (2007a: 123), for example, argues that the truth of statements concerning a disposition's
manifestation is always preserved under substitution of co-referring terms. Furthermore, Bird
questions Molnar's claim that dispositions are directed towards indeterminate manifestations,
and also suggests two further features of intentionality which dispositional directedness does not
share (see Bird 2007a: 118–26). These features are the extrinsicness of intentional states, and the
feature that the object of a thought is often the cause of that thought. See also Mumford (1999) for
further criticisms of the intentionality account of dispositionality.

should not be thought to consist in a genuine relation; he writes that 'the nexus between the intentional state and the object to which it refers is *not that of a genuine relation*' (2003: 62). A 'pseudo-relation' would therefore seem to be a better label for dispositional directedness. The problem with this concession, however, is that Molnar's account of dispositional directedness is then put in a similar boat to Heil's. On either account, we are left with a picture in which disposition tropes reach out in a rather mysterious way to their non-existent manifestations.

A route that some trope theorists might consider taking is to reject the 'pointing beyond' metaphor, and claim that the directedness of an unmanifested disposition trope is rooted in facts about how that trope resembles other tropes that *have* manifested in a certain way. This move comes with its own host of problems, however. To begin with, if facts about what a trope is a disposition for are not determined by the nature of the trope itself, but, rather, are inherited via resemblances to other things, it is hard to see in what sense that trope may any longer be said to be in and of itself powerful, as pan-dispositionalism requires. Further puzzling questions can also be raised about the resembling tropes which have manifested. What fixes *their* directedness, which is to say what fixes their identity as dispositions, prior to them being manifested? This strategy creates more difficulties than it solves. The trope theorists' directedness must, as Heil maintains, be 'built in' to the dispositions themselves, as difficult as that claim is to cash out.

Now that I have argued trope versions of pan-dispositionalism are at a significant disadvantage as compared with the universals account, I will conclude with some further comments concerning the details of the 'universals' account of irreducible dispositionality. In particular, I will focus on questions concerning the nature of the second-order manifestation relations which, on this view, account for the directedness (and so the identity) of dispositions.

9   The nature of second-order manifestation relations

I now want to briefly explore what it means to say that a relation—a second-order one in this case—is internal. We will see that there is more than one sense in which a relation can be internal. It will then be suggested that the sense of internality appropriate to the two-sided version of pan-dispositionalism, mentioned earlier, may be slightly different to the sense of internality appropriate to the dispositional monist version of pan-dispositionalism.

Famously, the notion of an internal relation, and the extent to which the world contains such relations, was strongly debated by British philosophers during the early part of the 20th century. Bradley, an idealist, says of an internal relation that it '...must at both ends *affect*, and pass into, the being of its terms'

(1893: 364). Joachim makes an equally elaborate comment when he writes that internal relations 'qualify or modify or make a difference to the terms between which they hold'(1906: 12).

Russell (1910: 160) tries to sum up what is common to all accounts of internal relations with the claim that internal relations (i.e., *all* relations, according to the British idealists), are 'grounded in the natures of the related terms'. This expression is itself somewhat vague, however, and Russell confesses to be uncertain about how, precisely, the expression 'natures of the related terms' is best understood in this context. In discussing this issue, Moore (1919: 62) offers two possible interpretations: either internal relations are grounded merely in the numerical identity of the terms themselves, or, more specifically, they are grounded in the 'qualities' the terms have, independently of their 'relational properties'. The crucial difference seems to be that, on the second view, essential reference is made to the intrinsic, qualitative natures of the things internally related, whereas on the first interpretation, internal relations are said merely to make a numerical difference to the terms related.

In fact, Moore thinks that, generally, those who speak of internal relations understand them in the second, stronger sense. It seems, however, that one could commit to the view that internal relations make a difference to the numerical identity of their terms, without committing to the further claim that those internal relations are grounded in the intrinsic qualities of their terms. In discussing the first, weaker formulation of internality, Moore specifies it in a precise form, using the symbolism of Russell and Whitehead's *Principia Mathematica*: 'The assertion with regard to a particular term A and a particular relational property ϕ, which A actually has, that ϕ is internal to A means then: $(x) \neg \phi x$. entails . $x \neq A$' (1919: 54). In words, this states that if some entity does not bear relational property ϕ, then that entity cannot be identical to A (given that ϕ is internal to A). This claim is also logically equivalent to the rather minimal claim that if x is identical to A, then x must bear relational property ϕ (given that ϕ is internal to A), i.e.: (x) $x = A$. entails . ϕx (1919: 54).

The question relevant for us is as follows. Are the pan-dispositionalists' second-order relations amongst universals internal in the weaker sense just described? The answer is clearly yes. It follows from Moore's definitions that A could not exist in any possible world without bearing relational feature ϕ (given that ϕ is internal to A). In other words, A bears relational property ϕ *necessarily*. This is, as we have seen, precisely the kind of claim that the pan-dispositionalists must endorse with respect to property universals. If the pan-dispositionalists were to deny that second-order manifestation relations amongst universals hold necessarily, thereby accepting contingency, they would ultimately be committing to the claim that properties are in and of themselves inert and categorical. On such a view, if a certain property were to bring any power at all to a world, it would do so only because a certain *contingent* relation (or relations) amongst universals happened to hold. The property itself would not be essentially or irreducibly dispositional. As mentioned earlier, this would leave one with a view closer to that of Armstrong (1997), which is clearly not a pan-dispositionalist view.

It seems, therefore, that at the very least, the pan-dispositionalists must accept the internality of (second-order) manifestation relations in the weaker sense defined by Moore. Recall, however, that a stronger sense of internality was also outlined. On the stronger interpretation, a relation is said to be internal if it is grounded specifically in the 'qualities' of its terms. In other words, if A bears relational property ϕ, and ϕ is internal in this sense, then if some entity x does not bear ϕ, then not only is x not identical to A, but x must be qualitatively different to A.

An important question, then, is this: are the second-order manifestation relations posited by the pan-dispositionalists internal in this stronger sense? In order to answer this question, one must ask whether it makes any sense for a pan-dispositionalist to speak of universals being *qualitative*. As was briefly indicated earlier, according to one version of pan-dispositionalism it does seem to make sense to speak in this way, whereas according to the other main version, it does not.

At the beginning of this paper two versions of pan-dispositionalism were distinguished: dispositional monism and the two-sided view. According to dispositional monism, the natures of all properties are exhausted by their dispositional characteristics, which means on the universals view under consideration that they are exhausted by their second-order manifestation relations. Thus, on this view, property universals are *wholly* relationally constituted, and so such universals can have no 'qualities' which could help to ground the second-order manifestation relations.[11] Therefore, if one advocates dispositional monism, second-order manifestations can only be internal in the weaker sense defined by Moore, which does not make reference to the qualities of the terms related.

The case is somewhat different with respect to the two-sided version of pan-dispositionalism. It is at this point that the main difference between the two-sided view and dispositional monism can once again be seen. According to the two-sided view, all properties have both an irreducibly dispositional aspect and a categorical (or 'qualitative') aspect. According to the view advo-

---

[11]The coherence of the thought that an entity may be wholly relationally constituted has been questioned by, for example, Heil (see 2003: 102–5, where Heil discusses Dipert's (1997) relationalist view about particulars, which is the first-order analogue of the dispositional monists' view about properties). Roughly, the objection is that if the 'relata' have no intrinsic features, then there is ultimately nothing there for the relations to relate, and so the picture collapses into nothingness. If this objection is fair, then this suggests the two-sided version of pan-dispositionalism is preferable to dispositional monism. It should be noted, however, that some philosophers argue, against Heil, that it is coherent to posit relata that have no features other than the relations they enter into. In order to avoid losing the relata, there merely has to be a mutual ontological dependence between the relata and relations, which means both that the relata do not exist independently of their relations and that the relations do not exist independently of their relata. This is the kind of view maintained by moderate structural realists; for further discussion of this kind of position see Esfeld and Lam (2008).

cated in this paper, the dispositional characteristics associated with properties are best understood as being rooted in second-order 'manifestation' relations amongst property universals. But unlike dispositional characteristics,
the qualitative aspects that two-sided theorists speak of are not rooted in relations amongst universals, because unlike dispositionality, categorical features
do not consist in directedness towards other properties.

Thus, if one is a two-sided theorist, then one will indeed think there are
qualitative aspects to properties which could potentially ground the internal
relations between universals in some sense. In order for there to be such
grounding, it would have to be the case that if a 'two-sided' universal bears a
certain internal relation (i.e., internal in the strong sense under consideration),
then if another universal does not bear such a relation, it must have a different qualitative side to the universal in question. In other words, there would
have to be a necessary connection between a property's 'qualitative' side and
its dispositional characteristics.

It should be highlighted, however, that in one place Martin, a two-sided
theorist, oddly leaves open the possibility that the dispositional aspects and
categorical aspects of a property may be merely contingently related (Armstrong et al. 1996: 87). If they were contingently related, then manifestation
relations would clearly not be internal in the sense under consideration. Martin's allowance of contingency has been heavily criticized, however. As Mumford (2007: 85) highlights when discussing Armstrong's dislike of the two-sided
view, if the categorical aspects of the world and the dispositional aspects of
the world really could exist apart, then the Martin-type position would look
more like a version of property dualism, in which case it would no longer be a
version of pan-dispositionalism.[12] To avoid this problem, the two-sided pan-
dispositionalist has to consider second-order manifestation relations to be internal in the stronger sense outlined by Moore, which is to say that qualitative
aspects and dispositional characteristics must be seen to be necessarily related. Not only would some universal x fail to be identical to universal A by
lacking a certain manifestation relation which A bears, if it did lack such a relation, then necessarily x would also bear a different qualitative aspect.

Finally, a further way of expressing the difference between the two-sided
view and dispositional monism is as follows. Armstrong has often characterized internal relations as 'ontological free lunches' in the sense that once
their relata, with all their qualitative features, exist, the relations are automatically there: they supervene upon their relata and so are 'no addition of being'
(1997: 12). An example of an ontological free lunch in this context is the resemblance relation: given the intrinsic natures of two objects, the nature of their
resemblance is automatically fixed. It can now be seen, then, that speaking
of second-order manifestation relations as 'ontological free lunches' is more
appropriate in the case of the two-sided view than the dispositional monist

[12] Place (in Armstrong et al. 1996) and Prior (1985) are two prominent proponents of categorical–
dispositional dualism.

view. If, as the dispositional monist maintains, second-order manifestation relations are what *wholly* constitute the nature of property universals, it would be inappropriate to class those relations as 'ontological free lunches'. One can hardly say that such relations supervene upon the prior natures of the universals, because they simply have no prior natures.[13] On the dispositional monist picture, therefore, internal second-order manifestation relations are quite different to many kinds of internal relation that philosophers speak of, such as the resemblance relation. Unlike the dispositional monists' internal manifestation relations, an entity's internal resemblance relations to other things can hardly be thought to *constitute* the entity in question. All of this suggests that when Armstrong speaks of internal relations as 'ontological free lunches', he has in mind those relations which satisfy the stronger definition of internality, discussed by Moore, which makes essential reference to the intrinsic 'qualities' of the related terms.

10   Summary

In this paper I have argued that if one is a pan-dispositionalist, one has special reasons for viewing properties as universals. More precisely, they will be universals which are, at least in part, relationally constituted. I have argued for this on the grounds that with universals in play, the pan-dispositionalist can satisfactorily account for the directedness of dispositions whilst at the same time respecting the fact that disposition instances are often intrinsic to their possessors and may exist even if their particular manifestations never come about. In contrast, it is less clear that a trope pan-dispositionalist can simultaneously account for all these facts in an adequate way. Finally, I briefly discussed the sense in which the pan-dispositionalists' second-order manifestation relations are internal on the universals view, and suggested that the internal relations to be utilized in the dispositional monist version of pan-dispositionalism are different to those figuring in the two-sided version.[14]

---

[13]This point has been emphasized by Barker (2009) in a recent discussion of dispositional monism.

[14]This research was supported by the AHRC Metaphysics of Science research project (2006–10). Versions of this paper were presented at the 2009 Metaphysics of Science Workshop in Granada, Spain (organized by Maria José Garcia Encinas) and the 2009 Metaphysics of Science International Conference in Melbourne, Australia (organized by Brian Ellis and Howard Sankey). My thanks go to the organizers of those events and to the audiences for their helpful comments.

THE CATEGORICAL–DISPOSITIONAL DISTINCTION

Sharon R. Ford

We have an overwhelming sense of the world as containing spatially-oriented distinct objects, and it seems that we derive this sensation from the properties of things as revealed by their effects upon us. This paper asks what sorts of properties should be posited to exist in accounting for this ostensibly qualitative, yet powerful, world.

My stance is a field-theoretic view, akin to Rom Harré's 'Great Field' (Harré 1970; Harré 2001; Harré and Madden 1975; Madden 1972), that describes the world as a single system comprised of pure power. Although it is outside the scope of this paper to detail such a pure-power ontology, here I defend the claim that structure should be considered in terms of pure-power rather than categoricity. This involves the further contention that 'pure-power' should not be interpreted as 'pure dispositionality', in the sense of potentiality, possibility or otherwise unmanifested power or ability bestowed upon a bearer. Rather, I view power as 'ontologically-robust', characterised in terms of *effect*, envisaging it as closer to Brian Ellis's notion of 'energy transmission' than to traditional ideas of dispositionality (see Ellis 2002 and his paper in this volume).

One major difference between my Foundation-Monism and Ellis's New Essentialism concerns the object level, at which Ellis assumes distinctness of objects and then builds in an ontologically-robust distinction between their categorical and dispositional properties. This distinction between property-types is posited in terms of passive versus active causal roles—what a property *is* versus what a property *does*. If categorical properties were ontologically-robust at higher levels, it would seem consistent that they obtain similarly at

181

fundamental levels. In this case, it would be reasonable to expect that such property-dualism involves fundamental categoricity.

In this paper, I argue that the causal role of properties appearing as active or passive is tied very closely to whether the relevant properties of an object are deemed intrinsic and essential or extrinsic and contingent, respectively (see note on p.199 regarding definition of 'intrinsic'). However, such a difference can occur only in ontologies whose objects are distinct. For those positing objects as merely supervenient—as non-distinct regions of a unified system— the distinction between intrinsic versus extrinsic properties borne by those regions, like the active versus passive and the categorical versus dispositional distinctions, does not apply in the ontologically-robust manner required for carving up reality into different property types. I suggest, instead, that these distinctions appear to arise at the object level by virtue of the assumptions built into the concept of object-hood; and stand as an instrumental, albeit useful, abstractions.

Applied to this view, consistency seems to demand that ontologically-robust categorical properties may not exist at higher levels if not at fundamental levels. Such elimination of categorical properties is supported by Strong Dispositional Essentialists, whose position is compatible with, but does not entail, a stronger claim that all properties are dispositional. I argue against this stronger claim because it appears to involve the dubious assumption that non-categoricity amounts to dispositionality. Distinguishing between 'pure power' and dispositionality, I contend that: i) there are no ontologically-robust categorical properties, although the appearance of them may be accounted for as higher-order and supervenient; ii) that the fundamental ingredients of the universe can be described in terms of pure-power, which is neither categorical nor dispositional; and iii) that the categorical-dispositional distinction arises only at the 'object level', although this is not in any case an ontologically-robust division of reality between two different natures in the sense described by Stephen Mumford (1998: 95).

1   Dimensions

Traditionally, both dispositional and categorical properties have been put forward in attempts to describe the manifest world. They have often, although not always, been defined in mutually exclusive and somewhat oppositional terms. Brian Ellis (2002: 68–10, 117, 174–5) notes that categorical properties have been considered readily imaginable; existing independently of behaviour; multi-dimensional; structural; non-dispositional or non-modal; and grounding or realising of the dispositional. Adopting the concept of quiddity, described by Alexander Bird and Robert Black as referring to some 'nature' of properties independent of their causal roles (Bird 2005a; Black 2000; Ellis this volume), Ellis asserts that categorical properties are quiddistic in the sense that they have their identity by virtue of what they *are* rather than by what they *do*. Categorical properties have by and large been characterised in terms of spatially-extended or space-occupying properties represented by Lockean primaries of size, shape, solidity and so on (Locke 1690: Bk II, Ch. VIII). Charlie Martin (1997), for example, describes qualitative properties as those needed for things to be perceived, providing the 'what' or 'shell' of objects; and John Heil (2007) describes them as what individuates or differentiates powers. Others describe their status as 'actual' or ontologically-robust (see Place in Armstrong et al. 1996: 49–67), or focus on their self-containment in terms of 'completeness' in their instantiation. David Armstrong describes them as 'exhausted' in their instantiation by particulars, whereby they do not reserve of themselves for further interactions with other particulars (Armstrong 1989: 118; Armstrong 1997: 41, 69, 245). Bird describes them as properties that have primitive identity (2007a: 45). Strong Categoricalists, such as Armstrong, hold that all properties, including those at the fundamental level, are categorical.

Dispositional properties have been contrasted with categorical properties in all of the descriptive contexts above. As Ellis (2002: 69–70,174–5) points out, dispositional properties have been considered uni-dimensional; non-structural; essentially modal; and grounds for the categorical. In this volume, Ellis defines dispositional properties as those that obtain their identity by virtue of what they dispose their bearers to *do*. Dualist positions, such as the New Essentialism advocated by Ellis (2001, 2002, 2008), hold that dispositional and categorical properties present a real difference in category between dispositional and categorical properties, whose mutual exclusion is based upon whether a property is structural or not (2002: 70).[1] For Ellis, although both types play causal roles (2001: 9–10; 2005b: 470), dispositional properties include causal powers, identified by the power that they bestow upon their bearers. In contrast, categorical properties—such as sizes, shapes and spatiotemporal relations and locations—obtain identity in virtue of what they *are* rather than what they *do*. Their causal role is passive, being that of merely 'factors' (Ellis 2010), rather than of driving forces, in the operation of causal processes.

Importantly for Ellis, categorical properties are structural properties, and he relies on these to underpin his natural-kinds hierarchy and its central tenet—that ontologically-robust structure is built into the universe (2001: 2; 2002: 68,174; 2005a: 382). This dependency on structure requires it to exist at fundamental levels and to include spatiotemporal relations, as he considers space and time to be 'the pure forms of physical structure' (Ellis 2002: 174). To accommodate current theories of physics, Ellis leaves open the idea that quantum fields might be fundamental quiddities, replacing the Lockean quiddities of Newtonian mechanics (2010). Yet, as he observes (2001: 10; 2008: 143), neither structure as relations between parts, nor objects themselves, exist at fundamental levels. In this volume, Ellis describes the distinction between dispositional and categorical properties *at the object level*. One reason is because it is only at this level that talk of 'physical objects that are capa-

---

[1] Dispositional essentialism is first posited jointly with Caroline Lierse (Ellis and Lierse 1994).

ble of having causal powers manifestly' (this volume) has meaning. Consequently, structure is portrayed by Ellis in two different ways: first, higher-order block structure (2001: 10, 247); and second, the kind of quiddities, regardless of whether Lockean or quantum mechanical, that feature as fundamental categorical properties (2001: 138, 218; 2002: 70).

Since they provide the structural component of the universe, New Essentialism's properties are almost all counted as dimensions. Ellis describes dimensions as the quantitative properties that are involved in the laws of nature, which direct how the effects of causal power are distributed (2001; 2008). The effect of a causal process is to change the values of certain dimensions, and Ellis (2008) describes these dimensions as 'respects in which things may be the same or different'. They 'determine the structural frameworks within which the powers operate' (Ellis, 2002: 174) and include, for example, quantities, shapes, duration, direction, spatiotemporal separation, position and time (Ellis 2001: 136–8; 2008). While the dimensions include most, if not all, of the categorical properties, they also include certain causal powers, since causal powers and capacities, like categorical dimensions, also represent 'respects in which things can be the same or different' (2008). I refer to the latter as 'powerful dimensions'.

The dimensions are 'determinables' (e.g. mass), each with at least two possible values or 'determinates' (e.g. 5 kg of mass), one of which is actual; and necessarily the dimensions are more ontologically fundamental than their values—mass must exist before one can have five kilograms of it. Importantly, the dimensions are not constituted by their values. Mass, for example, exists as something over and above the fact of being quantifiable. Ellis sees dimensions as 'among the fundamental constituents of reality at the object level' (this volume). However, rather than denoting dimensions as being actually fundamental, his specifying the 'object level' here leads me to interpret him as meaning that dimensions are *actually present* in the world in the sense of being ontologically-robust; his dimensions exist over and above their respectively many instantiated values.

The role of dimensions is to provide the circumstances in which the laws of action occur. How they do this is crucial to what the dimensions are. In his paper in this volume, Ellis provides the example of a weight suspended above ground level with the causal power to compress, stretch or pull things by virtue of potential energy. How the power manifests depends on the relevant circumstances, for example where the weight is in relation to other things, how it is fixed in position, and so on. The law of action concerns the effect of the weight in terms of the strength of the causal power as a function of the dimensions and initial circumstances. Ellis notes that all such laws of action are quantitative, depending on the magnitude and location of relevant powers; and that all involve one or more categorical properties, these categorical properties comprising the dimensions or circumstances in which the powers operate.

In this view, given that both dispositional and categorical properties may represent dimensions, being a dimension *per se* does not render a property categorical; being structural or quiddistic does. This raises the question of how a categorical dimension differs from a dispositional or powerful one *with respect to what role dimensions play* in general. For example, if the role of dimensions is to fix the circumstances for the action of laws by virtue of being 'passively' structural or quiddistic, then an 'active' dispositional property, such as mass, cannot be regarded as a dimension in these terms. And conversely, if dispositional properties are counted as dimensions, as Ellis suggests, then fixing the circumstances for the action of causal power is not the only role that dimensions play. In this case, properties that provide the structural circumstances cannot be deemed categorical merely on the basis that they are dimensional; and it is clearly not the fact of being dimensional that is crucial to Ellis's argument for the existence of categorical properties.

Ellis claims that categorical properties are quiddistic by virtue of being structural; such that they contribute to the circumstances for the operation of laws of action according to what they *are* rather than what they *do*. On these grounds, however, being a dimension *per se* does not justify counting structure as categorical rather than powerful. Instead, it is the claim for structure being *quiddistic* that purportedly renders it categorical.

2   The passive causal role of categorical dimensions

A difficulty—captured in the argument from quiddity—is encountered if the identity of a structural property is determined by means other than its causal role, since it would seem that, apart from some ability to engage in a causal process leading to our perception of such a property, we could not know anything about it. This problem appears to have been avoided by Ellis because his categorical dimensions *do* play a causal role (2005: 470), albeit a passive one. They are 'factors' in the causal process and, as circumstances for the operation of causal powers, feature in the laws of action that describe these powers.

This causal role, while allowing room for structural properties to be recognised by virtue of their relationship with causal powers, problematises the claim for categorical dimensions being purely quiddistic. As 'pure forms of physical structure', they have been described as restricting, constraining and informing the kinds of effects that causal powers can wield (Ellis 2002: 174). Yet Ellis denies that they produce any effect that can be attributed to their own action. They do not 'resist, deflect or otherwise interfere with the actions of any known causal powers' (this volume). While determining where causal powers may exist and how they are distributed, they do so not as causal powers or, if my earlier argument holds, even by being dimensions *per se*; but by sheer dint of existing 'structurally'.

Since categorical dimensions purportedly fulfil their causal roles without 'actively' acting, this raises the question of how we might know about properties that are deemed to *do* nothing. How might their effect-contribution, including their ability to affect our perception of them, be achieved? Ellis explains that we can know about these entities, not by virtue of their own abilities, but because of the abilities and actions of the relevant causal powers. He claims that 'the physical causal powers always act to change the values of the dimensions of the things on which they act' (2008), suggesting that the fundamental categorical properties might be discerned because they are respects in which things can change, and that this discernment is achieved by virtue of the causal powers of the objects that possess these categorical dimensions. Ellis (2001: 136–8) notes:

> Spatial properties, such as shape and size, are known to us because things of different shape or size affect us differentially. They produce in us different patterns of sensory stimulation, so that things of different shape and size look or feel different...But if spatial, temporal, and other primary properties and relationships are not causal powers, the question arises as to how we can know about them. We can know about them, we say, because of the dependence of the quantitative laws of action of the causal powers on these relationships. If the laws of action of the causal powers were independent of such factors as size, shape, direction, duration, spatio-temporal separation, and the like, then we could never know about them.

In a recent communication (unpublished manuscript, 2008), Ellis more explicitly describes and reaffirms how the categorical dimensions are discerned:

> The categorical dimensions of things are made manifest to us, not directly by their own powers, (for they have none), nor by our own innate capacity of perception (for nothing can perceive a quiddity directly), but by the distributed causal power of the things that possess them, and our innate capacity to learn from experience about the shape of this distribution.

This explanation requires a *relation* between the categorical dimensions and the causal powers of their bearers. In short, things possess categorical dimensions that change in response to the action of causal powers, and these changes are perceived and interpreted by us, allowing us to infer the presence of the categorical dimensions. This is in keeping with the idea, as expressed by Ellis (this volume), that 'the categorical properties of things can always be effective through the laws of action of the causal powers, even though they are not themselves causal powers'. Ellis goes on to describe distances and locations as factors in determining outcomes, although these are not causal powers. The upshot is that, in virtue of the causal role they play as they engage—via laws of action and reaction—with the causal powers, categorical dimensions escape the criticism that they are not discernible.

Earlier, I argued that being a dimension *per se* does not make a property categorical rather than dispositional. However, might being a dimension *per se* render a property powerful rather than categorical? The claim that quiddities can be known because of some causal role, even if passive, seems to raise the question of whether this role is fulfilled, not by virtue of the categorical dimensions being quiddistic, but because structure is powerful. The issue can be formulated in terms of teasing out what a property *is* from what a property *does*.

Let us suppose, as suggested by Ellis, that the role of categorical dimensions is to constrain and direct causal powers by limiting how they themselves can be changed. In this case, how changes can occur, and thus what the causal powers can do, seems 'built-in' to what the dimensions *are*. In this sense, the categorical dimensions are structural, yet play a causal, albeit passive, role. The problem is that the identity of these categorical properties is now determined not completely by what they *are*, but also by what they *do*. In this volume, Ellis suggests that we recognise at least certain categorical properties through common patterns of spatiotemporal relations, where these patterns are recognisable, not because of the categorical properties of the bearers, but because of essential dispositional properties that these bearers also possess. Thanks to these latter properties enacting patterns of behaviour upon the former, we discern the existence of categorical properties. However, I maintain that the categorical properties have *some* effect if the patterns arise by virtue of their interactive presence; and this appears to render them powerful *in some way*.

If 'quiddity' means what a property *is* over and above its causal role, then it can be argued that no such *pure* form of quiddity exists. However, rather than assuming the relevant property to be therefore dispositional, an alternative—and perhaps better—solution might be to recognise such properties as 'powerful'. Unlike the 'power-qualities' put forward by Charlie Martin and John Heil in their Identity Theory of Properties (Heil 2003; Armstrong et al. 1996; Martin 1997), however, I suggest that power is aligned with neither categoricity nor dispositionality. Properties that are powerful might be best described as those that both *are*, and yet *do*, merely by virtue of being. Although powerful, these may also be described as structural. Such properties are not categorical because they are not purely quiddistic, but I resist viewing them as dispositional because they are ontologically-robust, and hence always manifesting simply by virtue of existing. George Molnar proposes that we might allow such properties 'full ontological status on par with all of the paradigms of respectable existences' (Molnar 2003: 141). This requires powers to be more than mere possibilia, since it involves actuality in the sense of ontological-robustness.

3   Location

Ellis has argued that the passive causal role attributed to categorical dimensions as 'factors' in causal processes can be differentiated from the active causal role of causal powers. The difference between categorical properties and causal powers resides in this passive versus active distinction. It is tied to the fact of categorical properties having their identity in terms of what they *are*, while the identity of causal powers and other dispositional properties is given by what they *do*. However, if we cannot tease out what a property is from what it does, then the distinction between passive and active causal roles is compromised.

The spatiotemporal property of location (assuming it can be considered a property) is paradigmatically categorical for Ellis, and importantly, all causal powers are located. This is one reason why the laws of action describing causal processes all include categorical dimensions. In this section, I argue that seeing location as categorical depends not so much on the active-passive distinction, but more on whether the property is intrinsic to its object-bearer or represents an extrinsic relation between objects. (In this discussion, the arguments for relations between objects also apply to relations between mereological parts of complex objects).

The term 'location' is encompassed by the more general idea of spatiotemporal relations. Distance, for example, describes the location of one object with respect to another in terms of spatial separation. Location can also relate an object and a spacetime point within a frame of reference; even an absolute frame of reference, such as in the example cited by Ellis whereby a location, if emptied of all causal powers, would still be a location. Ellis notes that all instances of causal powers have specific locations, which are had contingently. He supplies the example of specific instances of gravitational mass. Objects are located here or there, but might have been located otherwise, or subsequently change their location through time. Thus the property of gravitational mass, although essential and intrinsic to those objects possessing it, is borne by objects that are nonetheless located contingently with respect to some spacetime frame of reference. By contrast, instances of location are said by Ellis to be necessarily where they are.

Location might be conferred with meaning in terms of the relation between an object and an absolute spacetime through which the metric for a fixed 'background' structure is presupposed (Kribs and Markopoulou 2005: 4). Alternatively, from a relationalist perspective, location has meaning only with reference to physical entities. On one hand, if location is derived via reference to a fixed background, then as Alexander Bird (2005a; 2007a: 161–8) and Stephen Mumford (2004: 188) suggest, this may constitute merely a choice of theoretical perspective. According to General Relativity, spacetime is not absolute, while various theoretical models in physics and in metaphysics treat spacetime as emergent (Bilson-Thompson et al. 2009; Bilson-Thompson et al.

2007; Harré 1970; (1975); Smolin 1997, 2000, 2006). On the other hand, if location is derived only with respect to the contents of the universe, then in keeping with the contingency of objects' locations, instances of location in general should also be deemed contingent. Moreover, even if it were hypothetically possible to remove all causal powers from a certain spacetime region, as Ellis suggests, the location itself would nonetheless be derived or 'framed' in relation to neighbouring objects.

Ellis attempts to base the categoricity of location on a contrast between the necessity of location instances and the contingency of causal power instances. The above argument amounts to proposing that this contrast only holds for theories that adopt a fixed spacetime background. I also earlier commented on why an active versus passive distinction between causal powers and categorical dimensions is unpersuasive. However, there is another option for an ontology that allows distinct objects to exist. This is to distinguish between Ellis's causal powers and categorical dimensions based upon whether the property or relation is intrinsic or extrinsic. I suggest that a more suitable criterion for location—and indeed, any spatiotemporal relation—being categorical, in Ellis's theory, is its being merely contingent in the operation of causal powers; and being contingent, I suggest, properly corresponds to its being extrinsic to the bearer of those causal powers. Taking distance, for example, Ellis argues that its being a factor in the outcome of a causal process—e.g. 'living a long way from Sydney prevents one from walking there'—does not itself constitute a causal power. We may think of causal powers in terms of dispositional properties, the possession of which bestows upon their bearers the ability to act in certain ways depending on the essential nature of these properties. Then distance does not bestow any *particular* ability (or in this case disability) upon the walker. Given that it features in many other causal laws, it is not an essential property or causal power intrinsic to the walker. Rather, since distance is extrinsic to the walker, it is a non-essential, categorical property, only contingently related to the walker. Nonetheless, it features in the law in question, playing a role in conjunction with other more specific powers that *are* intrinsic to the walker, such as endurance, muscular power and cardiovascular conditioning. Similarly, in the case of the weight example mentioned earlier; its *location* with respect to other objects contributes to the circumstances within which its causal powers operate, and which contribute to the laws of action of those powers. However, because its location is given with respect to other objects that are external to the weight itself, this cannot be an essential property of the weight, and hence cannot be a causal power of the weight. It must, instead, be a contingent, categorical factor that features in the laws of action specifying the effect of the weight.

Ellis defines a causal power as follows: 'Any quantitative property P that disposes its bearer S in certain circumstances C_0 to participate in a physical causal process, which has the effect $E - E_0$ in the circumstances C_0, where E is the actual outcome and E_0 is what the outcome would have been if P had not been operating' (this volume). A physical causal process is defined by Ellis to be an energy transfer from the state of one physical system to another, so as to bring about a physical change in the system that would not have occurred in the absence of that physical causal process. This description outlines two criteria for being a causal power: It must dispose its bearers to be involved in causal processes that i) involve transfer of energy; and ii) would thereby make a difference in outcome so long as the circumstances remain constant. (A causal process builds-in the idea that energy transfer occurs between states of different systems, although I see this as including energy transfer between different states of sub-systems of complex systems.)

Applying this description of causal power to the Sydney example: *Living a long way from Sydney disposes me, in certain circumstances, to not walk to Sydney, which alters the outcome of whether or not I go there.* No transfer of energy takes place, either actually or counterfactually, and so 'living a long way from Sydney', i.e. distance from Sydney, is not a causal power. But this example is complicated by being phrased negatively. Here is a parallel, but positive example: *Living close to a dairy disposes me to buy milk there rather than go without milk.* Here, energy is transferred in the process of walking to the dairy and bringing milk home. A physical change occurs, fulfilling one requirement for my location to be a causal power. The second requirement is that the presence or absence of closeness makes a difference to the outcome. Were I living further from the dairy, I would have no milk. So this criterion for distance being a causal power is also met.

According to the definition of a causal power, therefore, it is not clear from this example alone why distance should be treated as a *factor* or *circumstance* in my having milk rather than as a *causal power*. The reason could be that choosing to buy milk is typically seen as a contingent matter, even with the relevant circumstances in place. The supposed contingency is tied to the scenario whereby myself, my home, the dairy and the milk are all counted as distinct objects rather than constituting a single system. My buying milk is accordingly considered contingent by the Humean Principle of Independence, which disallows necessary relations between distinct objects (see Armstrong 2000: 8). Causal powers, however, do not represent contingency in this way. The power to crush an object under a weight, for example, would operate necessarily, providing that the circumstances specified in the laws of action of that power were in place. (I am putting aside probabilistic causal powers or propensities for the purpose of this example.) Thus, the claim that distance is merely a factor rather than a causal power depends on whether the separation between objects is viewed as extrinsic and contingent or as intrinsic and essential.

How might we observe distance as intrinsic rather than extrinsic? Suppose Mars and Venus were 50 million miles apart. The two masses could be counted as causal powers with specific locations (possibly identifiable with particular singularities in the gravitational field). But if the situation is viewed from a field-theoretical perspective, taking into account the entire region's contours of gravitational potential, then all the field contours throughout the 50 million-mile region are directly involved in how the whole set of field contours will behave—and this is the only relevant behaviour that will take place. Given any contingent variation in the local field contours, such as constituted by the presence of Earth, for instance, the field contours of the entire region would then behave differently. In fact, given the field topology at any time, and its subsequent behaviour, we could retro-determine the distance between Mars and Venus. (The topology and behaviour of the topology need not explicitly incorporate distance). This perspective implies that the relevant causal power (in this case gravitational power) is not really *at* any particularised location—but exists everywhere throughout the region of interest. At any time, any difference in the field contours of the whole region would cause different behaviour, and since distance is an intrinsic aspect of the field contours, it thus corresponds to causal power. Clearly, the distance could be retro-determined in various ways—in terms of other causal powers and laws—which might be taken by some theorists to suggest its ontic independence.[2] For example, the distance between any two charges could be retro-determined from the topology of the electromagnetic field at a given time, along with that field's subsequent behaviour. (Naturally, at this scale, the scenario is extraordinarily complicated by the phenomena of quantisation.) Both gravitational and electromagnetic scenarios will determine the same distance, but this simply indicates an intrinsically deep connection between the scenarios. It highlights that the singularities of gravitational fields could simultaneously be the singularities of electromagnetic fields. (Of course, it is commonly, if optimistically, anticipated that a unification of all the fundamental forces will be discovered.)

Perhaps being an extrinsic, contingent relation between distinct objects or states, versus being an intrinsic, essential property, amounts to the difference between what is deemed a categorical dimension and what constitutes a causal power. If so, then the difference is theory-bound rather than ontologically-robust. Say, for example, my home, the dairy and the land stretching between them are viewed as part of the same, very complex system rather than as distinct objects. In this case, just as in the Mars-Venus case, distance (as 'size') is an intrinsic property. Likewise, the weight, the wire and the object situated for compression by the weight may all be considered a single system to which the distance between the weight and object is intrinsic and essential.

4   Consequences

In a possible world containing just a single system—as on a field-theoretic view—extrinsicality, and hence contingency, is removed. An ontology that

---

[2]This was a point made to me by Brian Ellis in personal communication.

builds-in distinct objects requires contingency in the relations between such objects; ontologies that deny distinctness between parts have no need of categorical properties to supply contingency in the form of extrinsic relations between parts. The cost for this latter Foundation-Monist view, which I endorse, is that ontologically-robust possibility must also be denied. I think that denying categorical properties is defensible providing that possibility is accepted as merely an epistemological abstraction, which can be formulated in various ways. Taking a 4-dimensional block universe (4-D) perspective, for example, the intersections of object world-lines represent interactions. At any given time slice, the possibility of two world-lines intersecting is defined by the conjunction of their respective 'future light cones'. The extent to which light cones overlap pertains to the distribution of mass-energy associated with power or potentiality. However, the notion of 'possibility' embedded within our use of counterfactuals is pertinent because we are blind to the future. We do not possess a 'God's-eye point of view' to know 'the end from the beginning' (see: Isaiah 46: 10). Observing whether any two world-lines actually intersect, God has no use for possibilities. In a 4-D world-model, possibility arises due to the inability to see time slices 'ahead'. In a purely relational universe model, possibility would arise similarly, due to the inability to 'see' beyond a certain radius within any relational net. In either case, it is an epistemological abstraction.

Considerations of whether the universe is fundamentally indeterministic, with ontological possibility built-in as randomness, are presently under discussion in physics and philosophy. Issues include whether the probability that features in Quantum Mechanics is subjective or objective, and whether measurement entails irreducible uncertainty (see Caves et al. 2007). Regardless of these debates, it seems that the *appearance* of possibility can be linked to the emergence of ostensible higher-order categorical properties at the object level, where contingent relations between objects are called for. It is interesting that categorical properties and relations might actually provide for contingency, and thus for possibility, considering that dispositional properties are traditionally posited as the harbingers of possibility or potentiality.

One reason for distinguishing between dispositional and categorical properties only at the object level is, as Ellis notes, because this is where causal powers and their bearers exist. By providing the contingent circumstances for the manifestation of causal powers, categorical properties and relations are deemed responsible for accommodating unmanifested dispositional properties. In his paper in this volume, Ellis suggests viewing dispositional properties in two different ways: Unconditional (e.g. 'propensities') and conditional (e.g. causal powers) (this volume). For Ellis, the propensities are more primitive than the causal powers, and their laws of action are independent of contingent circumstances that involve categorical properties. An example might be a substance that undergoes spontaneous radioactive decay. In contrast, conditional dispositional properties, such as causal powers, rely upon categorical properties such as location and spatiotemporal relations to provide the circumstances in which their relevant laws of action and reaction operate.

It seems that only conditional dispositional properties, such as the causal powers of objects, require categorical properties for providing the contingency that makes them conditional. Hence there is good reason to suppose that the categorical-dispositional distinction goes hand in hand with accounts of distinct objects. This raises two questions: First, whether ontologies that do not incorporate categorical properties at the fundamental level can build them in at the object level in order to account for contingency; and second, whether ontologies that accommodate contingency, by positing ontologically-robust categorical properties at the object level, are constrained to also include them at the fundamental level.

Answering the first question, I think that if ontologically-robust categorical properties are absent at the fundamental level, then the best that can be achieved at higher levels is to account for categoricity in instrumental terms. One approach is to consider what we really mean by the units of dimensions, say, mass or distance.[3] Certain quantities in fundamental physics can be reduced to dimensionless numbers, dispensing with units altogether. Choice of units such as the second or the metre is often a matter of convenience and to a large extent reflects accepted physical theory. For example, as James Hartle notes, the second is defined as 'the time required for exactly 9,192,631,770 cycles in the transition between the two lowest energy states of a Cesium atom'. Employing the observation that the speed of light is the same in all inertial frames of reference, the metre is then defined as 1/299,792,458 of a second (Hartle 2003: 541). We have separate units for mass, length and time because our prior physical theories used independent standards for these quantities. The metre, for example, was defined by the distance between two marks on a particular bar, and the second was defined as a certain fraction of the mean solar day. Developments in physical theories, however, have come to show the interdependence of dimensions as measurements change to reflect updated information. As absolute quantities, both the speed of light and Planck's constant are frequently assigned the value of unity. The 'kilogram' has been traditionally defined as 'the mass of the block of metal kept in the Bureau International des Poids et Mesures, in Sèvres' (outside Paris). Today, as Hartle (2003: 542) notes, the kilogram can be defined in terms of distance: 'with confidence in the equality of gravitational and inertial mass, general relativity, and access to precise enough measurements, the kilogram could be defined as the mass of a sphere such that a test mass completes a circular orbit of radius 1 m in some defined number of days'. As Ellis notes in personal communication (Dec 2009), if it is in principle possible to measure distance, say, in terms of kilograms, then it might be also possible for all quantities to be measured in terms of dispositional ones. If so, then all quantities could be seen as derived dispositionally. This principle can be taken further to suppose that it might be

---

[3] I wish to thank Brian Ellis for this suggestion.

equally convenient (or inconvenient) to measure all quantities categorically, depending on instrumental purposes. This situation reinforces my claim that the categorical-dispositional divide is best viewed as a supervenient, higher-order distinction which, albeit intuitively appealing and instrumentally useful, embodies no ontologically-robust division of reality.

Coming to the second question posed above—should theories that posit ontologically-robust categorical properties, at the object level, also build them in at the fundamental level? Consistency would seem to demand an even-handed treatment whereby ontologically-robust categorical properties at the object level should derive from something more fundamental. This is the position that I believe applies to New Essentialism, although it is not clear how such categorical properties might be built-in. They are thought necessary in order to provide an account of structure, but if all categorical properties can be given in terms of dispositional ones, as argued above, then the burden of proof is on the dualist to show why structure should be considered categorical rather than powerful.

One important reason Ellis provides for rejecting the consideration of structure as powerful rather than categorical is couched in the form of a neo-Swinburne regress argument. The Swinburne regress objection is that a purely dispositional world ultimately lacks the resources to allow for the detection of properties and their effects. Richard Swinburne argues that a regress occurs for such worlds: We recognise powers by their effects, which are recognised in terms of the properties they involve. If these properties are themselves nothing but powers, then effects must be recognised by effects which must be recognised by effects, and so on; but at no stage are the required properties encountered. This regress can be broken, submits Swinburne (1980: 317), only if there is something more to properties than powers. The idea of structure being categorical is thus driven by calls for the effects of causal processes to be directly observable at some point, for which categorical properties are purported to be necessary; supposedly affording direct perception of effects. Ellis (2008) claims that, at some point in the causal chain, changes must occur in 'directly observable dimensions of things'. In this volume, he further notes that, although causal powers also give us direct knowledge of the world, quiddities 'are among the most direct objects of knowledge that we have of the world'. However, I suggest that almost the opposite seems to happen if, as noted earlier, these are observed only indirectly via patterns of distributed causal powers.

As I have argued, we are: i) not able to tease out what a property is from what it does; ii) not able to directly detect quiddities, which leaves the Swinburne regress unresolved; and iii) on the understanding that certain absolute physical quantities permit re-interpretations of measurement, not able to clearly differentiate between categorical and dispositional quantities. Accordingly, I contend that we should consider the dispositional-categorical distinction to be merely instrumental, and supervenient upon a pure-power world.

Resistance to a pure-power ontology seems to come from intuitions that the scientific picture concerning object-hood should mirror the ostensible objects of the manifest world, which would seem to entail fundamental particularity. However, in line with Michael Redhead, Paul Teller, Carlo Rovelli and others, I think that there are good reasons to deny the existence of fundamental particularity.

5   Rejection of Strict Particle-hood

There appears to be a prevalent, natural bias toward positing fundamental categoricity, connected with the intuitively attractive idea of fundamental particularity (i.e. haecceity or primitive thisness). Particles have traditionally involved 'primitive thisness' (Teller 1995: 29) or haecceity—described by Armstrong (1997: 109) as that which individuates particulars; a unique inner essence over and above any properties . Michael Redhead calls this 'transcendental individuality' (TI)—the idea of entities as 'bearers' of properties, or in Redhead's words, 'individuation that transcends the properties of an entity' (Redhead 1982: 59).

The reality of particles in modern physics and more recently in metaphysics is highly debated, with a consensus in favour of abandoning the traditional concept. Redhead argues that a traditional dualistic approach adopts two categories of entity: particles and forces between them. We may ask whether particles can be reduced to forces, and/or forces to particles. He shows that by quantising a field, we give it a particle aspect. In Quantum Field Theory (QFT), while particles are created and destroyed, they are, as Redhead (1982: 70) notes, 'just quantized excitations of particular modes of the field'. He likens them to the bumps in an active skipping rope, whereby quantisation does not entail that the field constitutes a collection of traditional particles. Redhead provides an extended argument why the distinction between 'field' and 'particle' can be tied to neither the distinction between boson and fermion, nor that between massless and massive fields (Redhead 1982: 72–6). Photons, for example, have zero rest mass, but because they carry energy and momentum, observes Redhead, we are inclined to treat them as substantial. However, he writes (1982: 80), 'it is not at all clear which is the "matter" particle and which is the "force" particle'.

Redhead's attempt to address the dilemma involves retaining the concept of particle while questioning the distinction between substance and force. He posits 'ephemerals', described in terms reminiscent of Lewis Carroll's 'Cheshire cat' as 'entities which can be distinguished one from another at any given instant of time, but unlike continuants cannot be reidentified as the same entity in virtue of TI at different times' (Redhead 1982: 88). Since a collection of indistinguishable particles may be described as an ephemeral in Redhead's view, this encompasses fields. He writes, 'like the Cheshire cat, although the substantial particularity has gone, there remains a particle "grin". The elementary "particles" are not particles but they are also not classical-type fields. They are quantum fields—ephemerals with a particle grin'. They are not classical fields, in this view, because they retain particle-like aspects such as energy and charge that come in discrete amounts (Teller 1982: 108). Thus, according to Redhead, the 'particle' and 'field' concepts are underdetermined in QFT.

Paul Teller adopts Redhead as a starting point, but takes his 'too soft treatment of ephemeral particles' further, to abandon any role for particularity. Teller claims that the notion of 'particle' in QFT is a relic of overlooking the fact that a full description, as per Feynman diagrams, must depict superposition of *all* processes mediating between input and output.[4] Partial or selective use renders the appearance of such diagrams as operating in terms of the particle concept to the exclusion of the superimposable field concept (Teller 1982: 109).

The argument for abandoning particles involves rejecting haecceity, or in Teller's preferred terminology, primitive thisness. He describes a hypothetical scenario whereby distinct particles (say, an electron and a proton) are distinguished by fixed, individuating properties. Teller argues that although it is 'natural' to read these fixed properties in terms of primitive thisness, attempts to formalise such a reading[5] lead to 'surplus structure', a term employed by Redhead (1975: 88) to formally describe elements that are absent in the 'real world' (Teller 1995: 20–6). That is, recognition of fixed properties will entail system components (e.g. non-symmetric vectors) that lack real-world counterparts. This failure of reduction from theory to the natural world is argued by Redhead to show that elementary particularity in the traditional qualitative sense does not exist. It represents what he calls 'one of the most profound revisions in our ultimate metaphysical *weltanschauung*, that has been engendered by our most fundamental physical theory, viz. quantum mechanics' (Teller 1995: 61–2).

Carlo Rovelli views particle-hood as a long-standing inference formulated in spite of the fact that the particle-aspect of these quantum 'entities' has never been detected and might be undetectable in principle. For Rovelli (1997: 191–2), particle-hood as traditionally conceived appears unsustainable:

> Indeed, a physical particle cannot be an extended rigid object, because rigid bodies are not admitted in the theory (they transmit information faster than light), nor can it be a pointlike massive object, because such objects too are incompatible with the theory (they disappear in their own black hole). Thus, understanding the physical picture of reality offered by general relativity in terms of particles moving on a curved geometry is misleading.

---

[4] A Feynman diagram symbolically represents sub-atomic 'particle' interactions according to all possible 'pathways'.

[5] For example, by using Labeled Tensor Product Hilbert Space Formalism (LTPHSF).

As Rovelli explains, fundamental particles could not be rigid bodies involving instantaneous transmission of effects from one side to the other, faster-than-light processes being ruled out by Special Relativity. Explicitly, the spacetime interval $Q^2 = c^2\Delta t^2 - \Delta s^2$ must be zero or positive for all physical processes. Thus, you cannot accommodate extension through space without sufficient accompanying extension through time, and any continuous, purely spatial extension is untenable (putting aside discussion of tachyons or pseudo-processes). Neither could particles be point-like: Given $F = Gm_1 m_2 / r^2$; particles with zero radius would form 'singularities' of infinite gravitational force, causally cut off from the universe.

John Gribbin (1998: 51–2) argues that the 'folk notion' of fundamental particles is basically a means to understand the mathematical laws describing fields of force, spacetime curvature and quantum uncertainty. Hence, electrons may be interpreted as 'energetic bits of the field' confined to a certain region (Gribbin 1998: 61) in accord with an application of Heisenberg's uncertainty principle relating energy and time. An energy variation multiplied by an associated time variation must be less than or equal to Planck's constant—determining the size of the quanta characterising field fluctuations. It is plausible that current physics theories will show that electrons, quarks and other candidates for fundamental particles can be properly understood solely in terms of underlying fields or microstructural topological arrangements of pre-space fundamentals that do not persist in the manner described of categorical entities or properties. It is feasible, therefore, to suppose that just as electrons, legitimately interpreted as charged regions of virtual photons, might boil down to manifestations of the field, so the categorical-dispositional distinction might be merely a higher-order distinction that is supervenient upon a pure-power base.

6   Summary and Conclusion

I have argued that the claim for categorical properties underpinning spacetime structure is not linked to their being dimensions; but rather, by virtue of their being quiddities. However, it is not clear that what these properties *are* can be teased out from what they *do*, compromising their status as quiddistic, where this means having identity by virtue of something over and above causal roles. The conflation of a property's *doing* with *being* also blurs the distinction between categorical dimensions as passive factors and causal powers as active drivers of causal processes.

Instead, the distinction between categorical dimensions and causal powers might be better understood in other terms; the former being extrinsic and contingent with respect to objects bearing causal powers, and the latter being intrinsic and essential properties of their bearers. Formulated in these terms, the distinction relies upon notions of intrinsicality and extrinsicality, which in turn presuppose the reality of distinct objects. I have suggested that this is why the categorical-dispositional distinction appears to arise at the object level.

Consistency would seem to demand that if such a distinction were ontologically-robust, heralding a real difference among types of properties, then it should obtain at fundamental levels, entailing fundamental categorical properties and, in turn, fundamental particularity. However, currently prominent theories of physics and some metaphysics seriously question the reality of fundamental particularity. Moreover, I have argued that location and other spatiotemporal relations must always be given either with respect to some fixed spacetime structure or by reference to the contents of the universe. The first way requires a background dependent ontology, the existence of which is under considerable doubt in modern physics and metaphysics. The second way should, I propose, treat location in the same manner as the objects in relation to which it is derived—as existing contingently. If this argument holds, then no real distinction can be drawn between location and causal power, based solely upon instances of causal power being contingent and instances of location being necessary.

I have also claimed that the consistent denial of fundamental categorical properties requires the categorical-dispositional distinction, at higher levels, to be merely instrumental and supervenient. This assertion is strengthened by noting that physics allows for conversion between measures of what have been seen as categorical quantities (e.g. distance) and dispositional ones (e.g. mass). The single-system approach of a Foundation-Monist theory requires no fundamental categorical properties. Such an ontology is consistent with a merely instrumental categorical-dispositional distinction rather than a reality divided into two types of property.

However, denying the ontologically-robust existence of categorical properties does not entail that all properties are dispositional. I point out a difference between what have been called dispositional properties and those that can be described as powerful in terms of ontologically-robust transmissions of effect. These properties can also be described as structural; they both *are* and *do*; but I do not view such pure-power properties as categorical, since they are not purely quiddistic.

I have tried to show that the distinction between categorical and dispositional properties is theory-bound rather than depicting an actual division into types of property. The cost of proposing a pure-power ontology is that an absence of contingency requires it to reduce the notion of possibility to an epistemological abstraction. This fee seems affordable compared to the inconvenience of incurring fundamental particularity. The benefits include being able to explain what intuitively appears to be a world comprised of fundamentally powerful structure. In contrast, dualist theories that build-in distinct objects, and therefore fundamental particularity, face problems in justifying the distinction between what have been considered to be categorical and dispositional properties, especially in light of modern developments in physics. Such a distinction renders structure purely quiddistic, but such quiddity would appear to be powerful merely by dint of existing. Accordingly, perhaps the distinction between categorical and dispositional properties should be re-evaluated in terms of a fresh notion of pure-power.

The push in physics is to find a unified theory integrating power and form. Whatever eventuates, current physical theories seem to herald a profound revision of spacetime, including location and other spatiotemporal properties and relations, in favour of these being emergent from something more fundamental, and conceivably purely powerful. I conclude that there is room for legitimate metaphysical scepticism concerning the existence of fundamental categorical properties, and concerning ontologically-robust distinctions that are predicated upon assuming the reality of distinct objects and fundamental particularity.

Note on the definition of 'intrinsic'

The concept of 'intrinsic' or 'intrinsicness' has been the focus of considerable debate in philosophical literature over the last three decades. In 1983 David Lewis separated the notions of 'intrinsic' and 'internal' on the basis that some properties can be only partially intrinsic. These include, for example, being a brother, being in debt, or being located with respect to some place. According to Lewis (1983: 197), properties that are entirely intrinsic (e.g. shape, charge or internal structure), are internal . This definition has been much discussed in papers on intrinsicness, with calls for a more precise delineation of properties and relations of objects that are co-relational with other objects. The formulation of the term 'intrinsic' that I use in this paper is primarily aligned with that provided by Robert Francescotti's (1999: 608) formal definition, given as follows:

> F is an intrinsic property $=_{df}$ necessarily, for any item x, if x has F, then there are internal properties $I_1, \ldots, I_n$ had by x, such that x's having F consists in x's having $I_1 \ldots, I_n$. (Call a property that is not a d-relational feature of item x an internal property of x.)

George Molnar (2003: 39) also gives an insightful definition of what it means to be intrinsic: 'intrinsic properties are those the having of which by an object in no way depends on what other objects exist' . Stated by him more formally, 'P is intrinsic to x iff x's having P, and x's lacking P, are independent of the existence, and the non-existence, of any contingent object wholly distinct from x' (2003: 102). The definition of intrinsicness that I will use in this paper incorporates key concepts that are central to all of these definitions: an intrinsic property is one possessed by an object which is, itself, not d-relational to any other distinct object. That is to say, as Molnar explains, an intrinsic property is had by an object independently of the existence of any other object. The terms 'd-relational' (i.e. relational to any distinct object) and 'independent from', in the above, are similar conceptually to Langton and Lewis's use of the term 'un-accompanied' or 'lonely' to discuss objects not contingently co-existing with other (distinct) objects (1998: 343). Adopting a compatible view of relations, I will use the term 'intrinsic relations' to refer to relations between properties of unaccompanied objects, providing that these relations may never differ be-tween duplicate pairs (i.e. pairs that have all of their internal properties the same). It follows that properties and relations that are not entirely intrinsic are in some degree external to their relata.

DISPOSITIONAL ESSENTIALISM AND THE LAWS OF NATURE

Barbara Vetter

1 Introduction

Dispositional Essentialism is the view that '[a]t least some sparse, fundamental properties have dispositional essences' (Bird 2007a: 45). This minimal characterization can be expanded in various ways, and various dispositional essentialists disagree on just how it should be expanded. Thus Alexander Bird holds, but Brian Ellis denies, that not only *some* but in fact *all* sparse, fundamental properties have dispositional essences; while Brian Ellis holds, and Alexander Bird is neutral on the question whether, the essentially dispositional properties are in turn essential to their bearers. In this paper, I will be concerned only with the minimal version of dispositional essentialism, and the argument that is given for it in Bird's recent book *Nature's Metaphysics* (Bird 2007a).

Dispositional essentialism is a form of scientific realism that has much to offer. The 'sparse, fundamental' properties are precisely those properties that participate in the (fundamental) laws of nature; but on the dispositional essentialist view, they do not only participate in laws. They *ground* those laws. Negative charge, for instance—if it is a fundamental property—is the disposition to repel other negative charges and attract positive ones; hence it is a law that negative charges repel other negative charges and attract positive ones. The same will hold, on the dispositional essentialist view, for other laws—at least the causal laws. Thus the laws discovered by such sciences as physics will not only be real, existing features of the world; they will be deeply rooted in the very fabric of the world. Bird begins his book with the observation that on the main competing views of laws, those of David Lewis and David Armstrong, it

201

would be very easy to imagine a world without laws. Ours could have been one of them. You can have all that it takes to build a world like ours—particulars possessing properties and undergoing processes in space and time—without laws; lawhood will be an additional, separate ingredient into the world. On the dispositional essentialist picture, if you take the laws away, there is nothing much left to make a world. For the dispositional essentialist, the tenets of scientific realism—that the laws discovered by science are a real, genuine part of our world—are not merely true, they are constitutive of what the world is made of.

Unsurprisingly, its ability to account for the laws of nature has been taken as the central advantage of dispostional essentialim. Ellis (2001) argues that dispositional essentialism can solve the three main problem with which rival views of the laws are confronted: the Necessity Problem (explaining the necessity involved in laws—the necessity expressed in the thought that negative charges *have to* repel each other); the Idealization Problem (explaining that many laws hold only *ceteris paribus*, i.e. can be interfered with); and the Ontological Problem (providing adequate ontological grounding for laws in reality). I have already indicated how a dispositional essentialist account of laws deals with the ontological problem; the idealization problem is addressed by the fact that a given law expresses the essence of some particular property, but in reality there will always be other properties operating, thus interfering through external forces. The Necessity Problem, finally, is solved very simply by the fact that laws are grounded in essences: like charges *have to* repel each other because that is their essence. The laws turn out to be metaphysically necessary. This is a controversial consequence of dispositional essentialism which I will not, however, be concerned with in this paper.

Intuitively, these are considerations in favour of dispositional essentialism. But they cannot fully come to bear until we have seen exactly how the dispositional essentialist view is spelled out, what exactly it is for a property to be essentially dispositional, and what the laws' being 'grounded' in those properties amounts to. Bird's book is notable for its careful attention to such questions, and for offering what is probably the most detailed formulation and the most extensive defence of dispositional essentialism to date. Where it goes wrong—and I will argue that it does go wrong—there is not, to my knowledge, any alternative available in the literature that would avoid the shortcomings. This is not to say that they cannot be fixed; but any fix will have to take Bird's formulation as a starting point and proceed from there.

Before I look at the argument for it, let me analyse Dispositional Essentialism in a little more detail. It can be understood in three simple steps:

First step: some fundamental properties have *essences*.

Second step: (some of) these essences are *dispositional*.

Third step: for a property P to be dispositional is for P to be characterized by a stimulus condition S and a manifestation condition M, which behave very much like a counterfactual conditional $Sx \,\square\!\!\rightarrow M x$ (with a *ceteris paribus* clause).

Note that nothing in any one step forces us to take the next one: one might hold that some properties have essences, but that these essences aren't dispositional; or one might hold that they are dispositional, but give a non-standard account of dispositionality. We will see, however, that all three steps are required for Bird's argument, and I believe that they are shared by everyone who commits themselves to dispositional essentialism.

In what follows, I shall understand 'Dispositional Essentialism' (DE) as the conjunction of these three steps. *Nature's Metaphysics* offers an indirect argument for DE. The indirect argument comes in two parts, one negative and one positive. The negative part aims to expose the explanatory poverty of DE's competitors, a cluster of views that Bird calls 'categoricalism'; the positive part then seeks to establish the superior explanatory power of DE itself. There is no attempt at a direct proof; rather, the overall argument is to establish DE as the best contender.

In this paper, I will argue that Bird's overall argument fails. I will do so not by disputing any particular step in the argument; I will grant that each one of these steps goes through. What I will dispute, rather, is that these steps establish DE as the best contender. I will argue that there are significant gaps in both the negative and the positive argument and that, unless these gaps can be filled, we have not been given a good reason to favour DE.

2   The Case against Categoricalism

The cluster of views that Bird opposes to DE is *categoricalism*. Categoricalism about properties is characterized as the view that (Bird 2007a: 67):

> [p]roperties are categorical in the following sense: they have no essential or other non-trivial modal character. For example, and in particular, properties do not, essentially or necessarily, have or confer any dispositional character or power. Being made of rubber confers elasticity on an object, but it does not do so necessarily. Being negatively charged confers on objects the power to repel other negatively charged objects, but not necessarily. In other possible worlds rubber objects are not elastic, negatively charged objects attract rather than repel each other.

Of the three steps towards DE that I set out in the preceding section, the categoricalist parts way at the very first step: for the categoricalist, properties do not have essences. It is a simple consequence of this, and not a further substantial categoricalist claim, that properties do not have *dispositional* essences. This is not to say that the categoricalist has to deny dispositional properties altogether: the characterization explicitly allows that properties

may 'confer' powers on objects. Is this not a 'non-trivial modal character'? It
is: powers, after all, are modal properties. But their modality is different from
that of essence or necessity. It is akin to the counterfactual conditional, or so
the third step towards DE contends, and the categoricalist has no quarrels with
that step. The first sentence, then, somewhat overstates the tenet of categoricalism. The categoricalist does not deny, qua categoricalist, that properties
have *any* non-trivial modal character. Rather, he denies that properties have
an essential or other necessity-like modal character. This, again, is merely to
say that the categoricalist denies only the first step towards DE.

Like DE, categoricalism is primarily a thesis about properties, but comes
with a view about laws. Unlike DE, categoricalism does not ground the laws
in the properties that participate in them: nothing about a property itself determines what these laws will be. The laws, therefore, have to be externally
imposed on the properties involved. Being externally imposed, they will plausibly be contingent: whatever it is that relates, say, charge and attractive force
in Coulomb's Law, having no special grounding in either the property of charge
or that of attractive force, could plausibly have failed to relate them. Again,
this is not so much a substantial further claim as it is a (plausible) consequence of the central categoricalist thesis about properties. It is not a logical
consequence, however: in a footnote, Bird notes that the categoricalist about
properties might in principle hold that the laws are necessary, but it is hard to
see a motivation for this on the categoricalist picture (Bird 2007a: 68, fn. 67).
The central tenet of categoricalism remains its denial that any property has an
essence or other non-trivial necessary feature.

If the laws are externally imposed on properties, what exactly is it that is
imposed? Bird considers the two main categoricalist answers to that question
and poses a dilemma for the categoricalist: on one answer, the Lewisian regularity theory, the laws fail to perform the explanatory task that is central to
lawhood; while the other, the Armstrongian view of 'nomic necessitation', is
forced to give up the central tenet of categoricalism.

According to the regularity view of laws, 'there is nothing metaphysically
deeper or more substantial from which laws flow. Rather laws are no more
than regularities that also meet some further, metaphysically innocuous, condition ... [for instance the] condition that laws are regularities that can be suitably systematized.' (Bird 2007a: 81f.) This, of course, is referring to the version
of the regularity theory famously proposed by David Lewis (for instance, in
Lewis 1994).

To understand what is at issue, it is useful to note (as Bird does) that DE
itself can be formulated as a regularity theory. According to DE, the laws are
those regularities that supervene on the dispositional nature of the fundamental properties. According to the Lewisian regularity theory, the laws are those
regularities that yield the best system. Both characterize the laws as a subset
of the regularities, a subset that meets a further condition. The crucial point is
the nature of that further condition: the Lewisian condition is to be 'metaphys-

ically innocuous', while the dispositional essentialist condition is anything but metaphysically innocuous. On the Lewisian view, not every regularity is a law, to be sure; but there is nothing special about the law-qualifying regularities themselves that sets them apart from others. Metaphysically, the laws are just regularities and no more.

It is this feature of the Lewisian view, and not anything more specific about the best-system account, that Bird finds unsatisfactory. His argument applies to any equally 'thin' theory. Let me use the term 'regularity theory' (as is usual) to refer only to the thin versions, excluding metaphysically 'thicker' accounts such as the version of DE just formulated. Bird's argument against the regularity theory is an elaboration of an argument first formulated by Armstrong (1983). It goes, roughly, as follows.

A first premise of the argument concerns explanation: nothing can explain itself. This is understood in a thoroughly metaphysical way: for p to explain q, the fact that p must be metaphysically something over and above the fact that q. (Bird substantiates this idea by talking about the 'ontological content' of a proposition.) A second premise concerns the role of laws: a law must explain its instances; or, if the law itself does not, then whatever it is grounded in must explain the instances. Thus if it is a law that all Fs are Gs, whatever that law metaphysically consists in or is grounded in must explain why a particular F is also a G.

According to the regularity theory, if it is a law that all Fs are Gs, then we can capture all there is (metaphysically) to that law by the universally quantified material implication $\forall x(Fx \rightarrow Gx)$. But this universal statement is nothing 'over and above' the particular instances—it is only, as it were, the collection of all of them. As such, it does not provide the explanatory resources to explain any one instance. The fact that a, which is an F, is also a G, cannot be explained by the further fact that b, which is an F, is also a G (nor that b and c, which are both F, are also both G, etc.); that latter fact is simply of no consequence to the former. Nor, of course, can it be explained by the one instance contained in the regularity that is of relevance—the fact that a, which is F, is also G; for that is the very fact to be explained, and nothing can explain itself.

If laws are to explain their instances—*all* of their instances—and regularities are nothing 'over and above' these instances, then the regularity cannot play the explanatory role of a law; rather, it is itself an explanandum for the law. Whatever is or grounds a law must explain, and so must be metaphysically something over and above, that regularity.[1]

---

[1] The Lewisian may dispute either the metaphysically 'heavy' conception of explanation, as being partial towards the metaphysically 'heavy' conception of the laws themselves; or he may point out that the best-system condition does, after all, have some kind of objective grounding in the world: they are not up to us, or so Lewis claims, cf. Lewis 1994: 232f. If this is not enough for 'over-and-above-ness', then again, the Lewisian may complain that the metaphysically heavy conception of over-and-above-ness is partial to the metaphysically heavy conception of laws. But I will press neither of these two responses; as I said, I am going to grant Bird all the steps in his arguments, to point out a lacuna that is left even if every step goes through.

If, as I am granting, the argument goes through, then it shows that laws need to be, or be grounded in, something metaphysically 'over and above' the fact that, say, all Fs are Gs. It is a further but overwhelmingly plausible assumption that this something, which distinguishes a law from a mere regularity, is somehow modal: if it is a law that all Fs are G, then not only will all Fs *actually* be G; they *have to* be G. This element of necessitation is what is missing from the regularity theory, and *it* is what is needed for a law to play its explanatory role: if Fs *have to* be Gs, then that modal fact is metaphysically 'thick' enough to explain why all Fs *are* Gs.

The second horn of Bird's dilemma considers the categoricalist view which accepts this and tries to balance the element of necessitation with the categoricalist commitment to non-modal properties. What is imposed on the properties that participate in laws, on this view, is a relation of 'nomic necessitation', or N for short, which holds between two properties F and G if, and only if, it is a law that all Fs are G. Nomic necessitation views have been defended by a number of authors, notably Armstrong (1983), but also Dretske and Tooley.

Bird agrees with the nomic necessitation theorist that an element of necessity is what is missing from the regularity theory, and that this is what laws need to explain their instances. He argues, however, that a categoricalist cannot have this element of necessity; that the combination of categoricalist contingentism and necessitation is not stable. To see why, we need to ask *how* it is that N necessitates, and thereby explains regularities.

Bird focusses on Armstrong (1983)'s version of the nomic necessitation view, and so will I. According to Armstrong, N is a second-order relation that holds between natural properties and has the notable feature of entailing regularity: $N(F, G)$ entails $\forall x(Fx \rightarrow Gx)$. That N holds between F and G is external to the latter two and is contingent; hence the nomic necessitation view is categoricalist. Yet *whenever* N holds, then so does the regularity; this is not a contingent fact, hence the element of necessitation is captured.

Has Armstrong succeeded in locating the element of necessity while upholding categoricalism? The properties F and G, to be sure, have been saved for categoricalism: they do not *have to* be related by N, nor do they have to be anything else of necessity or essence. But, Bird asks, what about N itself? Categoricalism is a claim about *all* properties. This should include relational and second-order properties, and in particular it should include the second-order relational property that is called nomic necessitation. But N is associated with an entailment; it could not hold between two universals without there also being a corresponding regularity. If N entails regularity, it thereby has what categoricalism denies to any property whatsoever: a non-trivial necessary or essential character.

This violation of the categoricalist doctrine is not a mere slip. If the relation between nomic necessitation and regularity were de-modalized and understood as mere material implication, then we would lose again any grasp on the modal element that it is supposed to possess 'over and above' the regularity. For N to relate universals with the right kind of modal force, N itself has to be related to regularity with some modal force. Nor will it help to apply to that modal force the same strategy that has been applied to the modal force of the laws themselves, and relate N to regularity by a third-order relation of nomic necessitation. That strategy, Bird argues, would lead into a vicious regress: the modal force will be continuously pushed up to ever higher-order relations of nomic necessitation, and at no point will we have succeeded in locating the aspect of necessity involved in a law. If that aspect is to be captured by nomic necessitation, it is through Armstrong's appeal to entailment—and that is, through giving up categoricalism.

I will, again, grant that the argument goes through: Armstrong, or indeed any categoricalist who wants to provide an explanatory role for laws at all, must reject categoricalism at least for the relation of nomic necessitation it-self. But Bird draws a stronger conclusion: Armstrong, he says (2007a: 92),

> must allow for there to be relations of metaphysical necessitation between distinct entities [namely, N and regularity]. But to do that is to permit potencies or potency-like entities. In which case [cat-egoricalism] is false. And furthermore, we may avail ourselves of these potencies or potency-like entities in explaining laws and re-pudiate the now-redundant N.

Now, *potencies* have been defined as 'properties that have dispositional essences' (Bird 2007a: 3). But nothing in the argument so far has forced Arm-strong, or anyone else, to accept that properties have *dispositional* essences. The friends of nomic necessitation may accept Bird's argument and yet be a long way from accepting DE. Armstrong, in particular, may retain his view that N entails regularity, and accept that this is to reject categoricalism as under-stood by Bird. But categoricalism as understood by Bird, while opposed to DE, is not its contradictory (i.e. the falsity of categoricalism does not entail the truth of DE). As we have seen, categoricalism parts way with DE at the first step. Armstrong, then, may be forced to take that first step and say that some property has a non-trivial modal feature, or an essence. Some property—that is, precisely one: N itself. Nothing in the argument we have been offered forces him to take any further step towards dispositional essentialism, or in particu-lar, to take the second step and accept potencies.

It may be argued that, once categoricalism is given up, there is no more reason to believe in nomic necessitation. For was N not a purely theoretical construct, introduced only to preserve categoricalism by reconciling it with the necessitating force of laws? But if it does not succeed in doing this, then, it may be said, why should we believe in such a construct in the first place?

But Armstrong will have various lines of response to that (rhetorical) ques-tion. Introducing N still preserves categoricalism for a wide variety of proper-ties: the first-order properties that scientists study. If we have to have some modality in the world, would we not rather have it located in one source?

Moreover, investing N with a necessary character preserves not only categoricalism for the first-order properties, it also preserves the contingency of laws.
If an essential or otherwise necessary character for one higher-order property
is what it takes to strike the right balance between necessitation and contingency, then that's a price worth paying, and true enough to the spirit (if not
the letter) of categoricalism to be far more attractive to the categoricalistically
inclined than the thoroughly modal world of DE.

Let me stress that I am not endorsing either side of this dialectic. My point
is merely that there is much argumentative space left between rejecting categoricalism and accepting DE. The negative argument has taken us a little way
towards DE (to be precise, one out of three steps). But that is not nearly enough
for it to give us a good reason to adopt DE, as Bird seems to think it does.

Bird's negative argument, then, shows less than it is intended to show. This
is not a devastating diagnosis: there is, after all, the positive argument. My
criticism of the negative argument only places greater weight on the positive
argument. It should give us very good reason to take not only the first step towards DE, but to go all the way, without bothering with the half-hearted intermediate views. So let me now look at the positive argument: DE's explanation
of the laws of nature.

3   The Explanatory Power of DE

Like Armstrong, the dispositional essentialist holds that the laws are something over and above the regularities. We have seen that DE can be phrased as
a form of metaphysically 'thick' regularity theory: the laws, according to that
version of DE, are those regularities that hold in virtue of, or supervene on, the
essentially dispositional nature of the sparse properties (Bird 2007a: 82). The
more general, official characterization of laws on the dispositional essentialist
picture comes later in the book (Bird 2007a: 201):

> (L) The laws of a domain are the fundamental, general explanatory
> relationships between kinds, quantities, and qualities of that do-
> main, that supervene upon the essential natures of those things.

This characterization provides an answer to the ontological question about
laws: they are grounded in the dispositional essences of properties. As we
have seen in the argument against the regularity theory, regularities are to be
explained by the laws, or by whatever grounds the laws. The dispositional
essences of properties, which ground the laws on this view, do explain regularities. Both the grounding of the laws, and the explanation of regularities, is
achieved in a most simple and elegant manner: by entailment.

According to DE, the fundamental properties at issue have dispositional
essences, characterized by a stimulus and a manifestation condition. Where
P is any fundamental property, dispositional essentialism says that P is essentially, and hence necessarily, a disposition to yield a particular manifestation

M in response to a particular stimulus S. Using as an approximation the conditional analysis of dispositions, this gives us:

(I) $\Box(Px \rightarrow (Sx \:\Box\!\!\rightarrow Mx))$,

which in a few simple steps of first-order modal logic with modus ponens for the counterfactual—assume $Px \wedge Sx$, derive Mx, discharge the assumption—leads to the statement of a nomic generalization:

(V) $\forall x((Px \wedge Sx) \rightarrow Mx)$.

(V) is a statement of a regularity; DE explains this regularity in terms of its derivation from (I). (V) is also a statement of a law, and we have a simple way of distinguishing a regularity that is a law from a regularity that is not a law: the former, but not the latter, can be derived from a (true) characterization of a dispositional essence, of the form of (I). The 'grounding' of laws in essentially dispositional properties has now a rather precise meaning: it can be cashed out in terms of logical entailment. Thus the Ontological Problem is answered: we know what the laws are grounded in, and what grounding consists in. The Necessity Problem is answered too: being derived from a necessary truth, (V) itself will be metaphysically necessary. The Idealization Problem, finally, will be addressed by the fact that the conditional analysis employed in (I) is only an approximation and holds *ceteris paribus*. (Bird, however, believes that this does not apply at the fundamental level.)

Presented in this abstract manner, the account is highly attractive. But let us fill in the schema and look at an example. One good candidate for an essentially dispositional fundamental property is charge—one of the favourite examples of dispositional essentialists. The law that characterizes charge, or rather: that should be grounded in charge, is Coulomb's Law. Coulomb's Law states the relation between any given determinate charge Q and any other charge q_i, the distance r_i between Q and q_i, and the attractive or repulsive force F_i that is exerted:

(CL) $F_i = \epsilon \dfrac{Qq_i}{r_i^2}$

Clearly, (CL) does not look anything like (V). A first and minor point of dissimilarity is that (V) is, and (CL) is not, stated in the form of a conditional. What is more important is that (CL) states a *function*, and the variables in it range over *quantities*. Charge, force, and distance are quantities: they are determinable properties that come with an ordered range of determinates. An object is not merely charged (positively or negatively), it has a particular determinate charge, say, charge e or charge $2e$. The same holds of exerting a force and being at a distance from something. Coulomb's Law, accordingly, states not merely *that* an object with a (positive or negative) charge will manifest a certain kind of force (attractive or repulsive) in response to a certain kind of stimulus condition (say, another charge at some distance from it); it states exactly *how much* force the object will exert in response to exactly *how much* charge at *how great* a distance.

(I) and (V), on the other hand, appear rather to be designed for qualitative properties. They can tell us merely that *if* such-and-such properties are instantiated, *then* such-and-such other properties will be instantiated too. But as Bigelow and Pargetter (1988) note, with quantities 'the simple "on" or "off" of being instantiated or not being instantiated seems to leave something out' (Bigelow and Pargetter 1988: 287). Can the quantitative nature of charge and Coulomb's Law be integrated into the derivation of (V) from (I)?

To see how this might be done, we need to fix the minor dissimilarity and formulate Coulomb's Law in the form of a conditional. Now, (CL) is a function with several variables, and we do not want any free variables in a (V)-like statement of a law. There are two things we can do with the free variables: we can fill in determinate values for them, or we can quantify over them. For simplicity's sake, let us focus on a particular determinate charge, say, electric charge (charge e). Then the first strategy yields an infinity of rather specific statements, one of which is:

> (V-1) $\forall x$ ((x has charge $e \wedge x$ is 5.3 $\times 10^{-11}$m from a charge of 1.6 $\times 10^{-19}C$) $\rightarrow x$ exerts a force of 8 $\times 10^{-8}$N).

The second strategy, on the other hand, yields only one, multiply quantified, statement:

> (V-$\forall$) $\forall x \forall r_i \forall q_i$ ((x has charge $e \wedge$ x is at a distance of r_i from a charge q_i) $\rightarrow x$ exerts a force of $F_i = \epsilon \frac{eq_i}{r_i^2}$).

Note that (V-$\forall$) is closer to Coulomb's Law, but (because of its multiple quantification) is not quite of the same form as (V). (V-1), on the other hand, is an instance of (V). Accordingly, (V-1) can be derived from an instance of (I):

> (I-1) $\square$ (x has charge $e \rightarrow$ (x is 5.3 $\times 10^{-11}$m from a charge of 1.6 $\times 10^{-19}C$ $\square\!\!\rightarrow x$ exerts a force of 8 $\times 10^{-8}$N)).

(V-$\forall$), on the other hand, can be derived from the following:

> (I-$\forall$) $\square$ (x has charge $e \rightarrow \forall$ charges $q_i \forall$ distances r_i (x is at r_i from $q_i \square\!\!\rightarrow x$ exerts force $F_i = \epsilon \frac{eq_i}{r_i^2}$))

Again, (I-1), while an instance of (I), is not a characterization of charge e; and (I-$\forall$), while being a better candidate for characterizing the dispositional essence of electric charge, is not quite of the same form as (I). There is a tension here: we can either keep strictly to the form of Bird's derivation, or we can capture the law (Coulomb's Law) and the property (electric charge) that we intuitively want to capture; but we cannot do both. I will argue that this tension is no mere appearance: there is a real conflict here, and it cannot be resolved without giving up some part of DE.[2]

---

[2] Alice Drewery has independently raised worries very similar to the ones I am going to discuss, in a talk given at the Metaphysics of Science conference in Nottingham, September 2009.

The question that I put to the dispositional essentialist, then, is this: which is the fundamental property, and which is the law to be derived: (I-1) and (V-1), or (I-∀) and (V-∀)?

Given his explicit preference for the property of electric charge (Bird 2007: 44), and the known quantitative nature of the properties and laws of physics, it is surprising that Bird does not devote a substantial part of his exposition to this question. His answer to it is hidden in an early and somewhat peripheral section on multi-track dispositions (2007: 21–4), and is curiously forgotten in the rest of the book. In the section on multi-track dispositions, Bird argues for the priority of (I-1) and (V-1) over (I-∀) and (V-∀). Here is what he says (slightly rephrased to fit my set-up).

(I-∀) is equivalent to a conjunction of (infinitely many) statements such as (I-1). Undoubtedly (I-1) characterizes *some* disposition; the question then is, which of them is more fundamental, (I-1) and its cognates, or (I-∀)?

Bird (2007a: 22) defines a 'pure disposition' as one 'which can, in principle, be characterized … as a relation between a stimulus and a manifestation'. (I-1) gives the characterization of a pure disposition; (I-∀), as we have seen, is equivalent to a conjunction of pure dispositions. And Bird continues:

> It is my view that all impure dispositions are non-fundamental. Fundamental properties cannot be impure dispositions, since such dispositions are really conjunctions of pure dispositions, in which case it would be the conjuncts that are closer to being fundamental.

The fundamental properties are those characterized by (I-1) and its many cognates, and not properties such as charge e, which is better characterized by (I-∀). It turns out not only that charge, or even charge e, is not, after all, one of the fundamental properties. It also turns out that Coulomb's Law, or even (V-∀), is not, after all, a law that can be derived from the dispositional essences of the fundamental properties. What can be derived are infinitely many law-like statements such as (V-1).

What, then, is the status of Coulomb's Law, or even the instance of it expressed in (V-∀)?

(V-∀) certainly states a regularity; but does it, on the present account, state a law? We have seen that a regularity will qualify as a law just in case it is entailed by the dispositional essence of a fundamental property, as expressed in (I). But there is no such dispositional essence to entail (V-∀); (I-∀), which does entail it, does not state the dispositional essence of any property. So (V-∀) does not seem to state a law. Similarly, there is no dispositional essence to entail Coulomb's Law in its full generality; so Coulomb's Law is not a law.

The same point can be made the other way around: assuming that Coulomb's Law, or at least (V-∀), is a law, what is this law grounded in? By the explication of grounding, it would have to be electric charge, i.e. the disposition expressed in (I-∀); but that has a rather questionable ontological standing.

If the property expressed by (I-∀) is 'really' just a conjunction, as Bird tells us, then it is 'really' nothing but its conjuncts. It is, metaphysically speaking, nothing over and above the conjuncts, in the same way in which a regularity has been argued to be nothing over and above its instances. But if it is nothing over and above the conjuncts, then it cannot provide the ontological grounding for anything over and above that for which its conjuncts provide the grounding.

If regularities themselves are explananda, then dispositional essentialism has not delivered the explanation it has promised. Its laws, (V-1) and its cognates, can explain *some* regularities—the regularity, for instance, that a given object exerts a force of 8×10^{-8}N whenever it is 5.3×10^{-11}m from a charge of 1.6×10^{-19}C. But the crucial regularity is the similarity *between* these specific regularities: the fact that they all exhibit the same mathematical correlations between stimulus and manifestation.

How is that regularity to be explained? It is clear that the 'impure disposition' (I-∀) will have to play a role. And indeed Bird says that '[w]hile it is possible to gerrymander impure dispositions of all sorts, it is clear as regards the cases we are interested in, [such as] charge [...], that the conjunctions are natural or non-accidental' (2007: 22). That is clear indeed; but it is far less clear how the dispositional essentialist is to account for the naturalness of these conjunctions. It will not do to say that (V-∀) is grounded in a non-fundamental disposition captured by (I-∀). For as we have seen, that non-fundamental disposition could not be, and hence could not ground, anything over and above the fundamental conjuncts of which it is made up.

If a single determinate charge such as charge *e* and the instance of Coulomb's Law that is expressed in (V-∀) cannot be captured by DE, then even less will the determinable property charge and Coulomb's Law be captured. So I conclude that, with Bird's preference for (I-1) over (I-∀), DE fails to accomplish its explanatory task when it comes to the property of charge and Coulomb's Law. To see just how damaging this conclusion is, we need to consider (1) how far the argument I have given generalizes beyond the one property and law that I have been considering; and (2) whether the competing views, the regularity theory and the nomic necessitation view, run into similar problems.

First: how far does the argument generalize? The answer is: very widely. It may be, in fact it is rather likely, that charge and Coulomb's Law are not fundamental. But my argument did not turn on any specific features of charge, or of Coulomb's Law. It turned only on charge being a quantity, and Coulomb's Law being a functional law that states mathematical correlations between quantities. Every candidate fundamental property that participates in a functional law will be subject to the same line of argument. In his essay for this volume, Bird conjectures, I believe for the reason stated, that the really fundamental laws will not have any constants, such as ε in Coulomb's Law. But not only is that an undesirably strong prediction; it will not be enough to avoid the kind of argument I have given. The argument applies to properties and laws that involve *any* mathematical operation—multiplication or division is enough. And Bird should certainly not predict that there will be no mathematical operations whatsoever in the fundamental laws. That would not only be a daringly strong prediction. It would make it utterly mysterious what *could* count as a fundamental law at all.

Second: how do the rival views fare? The answer is: better. The regularity theory can take regularities of the form expressed in (V-1), note the similarity between them, and recognize it as just another regularity. The nomic necessitation theorist can see regularities of the form expressed in (V-1) as grounded in a relation of (second-order) nomic necessitation holding between the determinate quantities, note the similarity between them, and see that further regularity as grounded in a (third-order) relation of nomic necessitation holding between the former (second-order) relations. (A strategy roughly along these lines seems in fact to be suggested in Armstrong 1983: 113.) In general, the categoricalist views connect the properties participating in a law by imposing the laws on them, as it were, from above. There is no problem in principle with going up one step further and applying the same operation again. DE, on the other hand, does not impose connections from above; it finds them all on the level of the properties themselves. Connections that are not provided for on that level cannot be found elsewhere, nor can they be imposed from above without giving up the central tenet of DE.

The answers to both of my questions, then, confirm that Bird's argument for the fundamentality of (I-1) rather than (I-∀) has very damaging consequences for DE. The natural solution is to give up the multi-track view and adopt (I-∀) as fundamental instead. That, of course, means that we must reject Bird's argument. Let us see how that can be done.

Bird said that dispositional properties such as the one expressed in (I-∀), or electric charge, are 'really conjunctions of pure dispositions' (2007: 22). Now, what is the status of that 'really'? Earlier, I compared the situation to the argument against the regularity view: there too the universally quantified material implication $\forall x(Fx \rightarrow Gx)$ was 'really just' an infinite conjunction, it was 'nothing over and above' its instances. But DE's solution to this problem was precisely to stipulate a property which grounded the regularity. Why should we not apply the same strategy to the conjunction that is now at issue?

The answer lies in what I have called the third step towards DE: the characterization of what it is for a property to be dispositional. Dispositionality, we have seen, is understood as connecting a stimulus property and a manifestation property, the connection amounting to something like a counterfactual conditional (with a *ceteris paribus* clause). The 'pure' dispositions are pure simply in that they perfectly conform to that characterization.

If we look at (I-∀), however, the characterization of dispositionality fails: we cannot take it apart in the way we can take apart (I-1), factoring it into two separate properties. With (I-1), the stimulus condition is: being 5.3×10^{-11}m from a charge of 1.6×10^{-19}C; the manifestation condition is: exerting a force of 8×10^{-8}N. In (I-∀), we have quantifiers ranging over q_i and r_i in both the stimulus and the manifestation condition. If we try to separate them and specify one in separation from the other, we lose the very correlation which (I-∀) has been formulated to capture. But not only can the properties not be separated from each other; the counterfactual conditional is not doing any work. What (I-∀) says is that everything with charge e will always exert a force that stands in a certain mathematical correlation to whatever other charges are present and their distance from it.

We can see now why Bird went for his multi-track view. It is the third step towards DE, the characterization of dispositionality, that forced it on him. If we think that a disposition has to come with separable stimulus and manifestation conditions, related in a counterfactual-like way, then the best (I-∀) can hope for is to count as a conjunction of such dispositions. If (I-∀) is to stand on its own, an entirely different conception of dispositionality will be needed: to begin with, dispositionality will look more like a one-place operator, but one that takes complex functions as its complements.

It is, of course, not an idiosyncrasy of Bird's to rely on the conception of dispositionality that he does rely on. This is the one established conception of dispositionality throughout the literature. Much of the recent debate on dispositions is taken up by objections to, and variations on, the 'conditional analysis' of dispositions—assuming that, if dispositions are to be analysed in any way, it will have to be in terms of a counterfactual conditional. Taking (I-∀) and such-like properties to stand on their own will require a radical departure from these extant views of dispositionality, and a considerable amount of work on the alternative view.

For instance, an account will be needed of how electric charge, the dispositional property characterized by (I-∀), is related to such specific sub-dispositions as (I-1) if it is to be more than just a conjunction of them. Note that, while the apparatus of determinables and determinates may enter in some way or other, it cannot be used to characterize the relation between electric charge and the many sub-dispositions. For the determinates of one determinable exclude each other: while having charge entails having some determinate charge, having any one determinate charge entails not having any other determinate charge. In the case at hand, however, having electric charge entails not merely having *some*, but indeed having *all* the many parallel sub-dispositions.

Another question that will need to be answered is just what kind of modality it is that is involved in dispositions, if it is not that of the counterfactual conditional. Is a one-place dispositionality operator to be understood in terms of plain necessity? And how does this characterization relate to the mathematical functions in its scope, and to the essentiality or necessity operator in whose scope the disposition is characterized in (I-∀)?

If, as I have argued, the third step towards DE has to be given up, the dispositional essentialist will have to provide a novel metaphysics of dispositions.

I am optimistic that such a view can be developed. In the meantime, however, the superior explanatory power of dispositional essentialism is yet to be demonstrated.[3]

[3] For helpful comments and discussion, I would like to thank Alexander Bird, Natalja Deng, Alice Drewery, Antony Eagle, Marc Lange, Gail Leckie, Markus Schrenk, and Timothy Williamson, as well as the audience at the Metaphysics of Science conference in Nottingham in September 2009.

Part VI

NATURAL KINDS

THE METAPHYSICS OF
DETERMINABLE KINDS

Emma Tobin

Armstrong (1997: 65–8, 1999) argues that all natural kinds supervene upon more simple monadic and relational universals. Moreover, since kinds supervene, then they are not an ontological addition to their base. In contrast, Ellis (1999; 2001; 2002) argues that natural kinds find a *sui generis* place in our ontology. Natural kinds require a distinct ontological category and are irreducible. Both Armstrong and Ellis agree that our scientific worldview should inform our general ontology. One should not proceed *a priori* to posit an ontological theory and then seek to reconcile it with the scientific worldview. Thus, the correct ontological account of natural kinds should be scientifically informed. In this paper, I argue that science suggests a more complex view of natural kinds than has been permitted by either of the above ontologies.

In Section 1 and Section 2, I discuss a disagreement between Armstrong and Ellis on the relationship between existential entailment and ontological dependence for natural kinds, which results in Armstrong's natural kind eliminativism and Ellis's natural kind essentialism. Section 3 argues that there is an important distinction between different kinds of determinable kind. Some determinable natural kinds have members that are determinates e.g. chemical elements such as uranium with its various isotopes). Other determinable kinds have members that are themselves complex determinables (such as *metals, species, proteins, enzymes* and so on). I show that at least some cases of determinable kinds (DET1, DET2 and DET3) can be accommodated by natural kind essentialism. However, in Section 4, I argue that a distinct kind of crosscutting determinable kind cannot be accommodated by Ellis's essentialist account (DET4). In particular, I argue that these kind categories crosscut each other in ontologically significant ways. Our metaphysics of natural kinds must allow for such cases of crosscutting if it is to be reconciled with the kind categories that we find in science. In Section 5, I nevertheless claim that re-

alism about DET4 kinds is the most plausible ontology that is consistent with
the scientific kinds which are found in natural science.

1   Supervenience and Natural Kind Eliminativism

Armstrong (1997) argues that a distinct ontological category is not required for
natural kinds. Kinds supervene on simple monadic and relational universals.
He states (1997: 67–8):

> It is of course a great fact about the world that it contains kinds of
> things ... The kinds mark true joints in nature. But it is not clear
> that we require an independent and irreducible category of uni-
> versal to accommodate the kinds. ... All the kinds of thing that
> there are, supervene. And if they supervene they are not an onto-
> logical addition to their base.

For Armstrong, determinate kinds can be reduced to determinate property universals. For example, the determinate kind *electron* can be reduced to
its determinate properties; namely mass, charge and the absolute value of the
spin. These are identical properties in all electrons. The kind electron is a simple conjunction of the three property universals (mass, charge and spin). Armstrong argues that conjunctive properties are not properties over and above
their conjuncts. Thus, all higher-level kinds are reducible to their constitutive
properties.

Armstrong (1997: 65–8) denies that there are any determinable universals[1]
and thus he denies that there are determinable kinds.[2] He states (1997: 50):

---

[1]Armstrong is committed to a sparse theory of properties, which does not allow a universal for
every meaningful predicate. For example, he denies negative and disjunctive universals, though
he does accept, as we saw above, conjunctive universals.

[2]Armstrong (1997: 246–8) later concedes that the outright denial of determinable universals
is too strong a position to take. He states that determinables are required for the variables in
functional laws. Consider the law of gravitation:

$$F = G\frac{m_1 m_2}{r^2}.$$

F, m_1, m_2, and r are determinables. But, if specific values for force, the two masses and distance are substituted into the equation then the law becomes a determinate of the original determinable. The law is a second order universal whose instances are determinate monadic universals
governed by a first order law of nature. The problem is that it is the law itself formulated as a determinable, which we consider to be the law of nature is this example. Armstrong agrees that we
would not want an indefinite multitude of laws for each different value of the variables linked in
the law. Thus, we must allow that functional laws are determinables. However, we do not need to
accept that laws are themselves natural kinds. Thus, allowing determinable functional laws is not
tantamount to admitting determinable kinds into our ontology. There are no determinable kinds
according to Armstrong. Equally, since determinate kinds supervene on determinate universals,
there are no determinate kinds either. Thus, there is no requirement for a natural kind category,
according to Armstrong.

> Because determinates entail the corresponding determinable, the determinable supervenes on the determinates, and so, apparently, is not something more than the determinates.

The existence of the determinate entails the existence of the determinable, but the existence of the determinable does not entail the existence of the determinate. Armstrong argues that if we know that Bs exist and that the existence of Bs entails the existence of As, then we add nothing new by postulating the existence of As over and above the existence of Bs. Thus, methane molecules are existentially entailed by the carbon and hydrogen atoms that comprise them. The direction of ontological dependence from the determinable (methane molecule) to the determinate (carbon and hydrogen atoms) is contrary to the direction of existential entailment from the determinate to the determinable.

Nevertheless, it is important to point out that though Armstrong is reductionist about the kind category itself, his account of higher-order universals is a supervenience thesis, where the higher-level complexity supervenes on simple states of affairs. In contrast, for reductionists, all higher-level kinds and properties are (either type or token) identical with underlying kinds and properties and thus, can be reduced to them. Thus, for the reductionist, for example, serotonin molecules existentially entail the carbon, hydrogen, oxygen, and nitrogen atoms that compose them. Yet, carbon and the other atoms can certainly exist without serotonin.

Pace Armstrong, for reductionists, existential entailment is identical to ontological dependence. Thus, serotonin is ontologically dependent on carbon, hydrogen, oxygen and nitrogen. Serotonin can at least, in principle, be reduced to the atoms that compose it. Moreover, these constituents can themselves be reduced to the more fundamental particles that compose them (electrons, protons and neutrons). Thus for reductionists, the determinable existentially entails the determinates and thus the determinable is ontologically dependent on the determinate.

2   Natural Kind Essentialism

As we saw above, Armstrong denies that the direction of ontological dependence is identical to the direction of existential entailment. In contrast, Ellis agrees with the microreductionist that the direction of existential entailment is identical to the existence of ontological dependence. However, rather than accepting microreductionism, he argues that this identity entails anti-reductionism. For Ellis, the determinable natural kinds (which Ellis calls generic kinds[3]) in every category are ontologically more fundamental than any of their species. For Ellis, determinable natural kinds could exist even if none

---

[3] I will refer to his generic kinds as determinable kinds to tie in with what is said in the remainder of the paper.

of their species existed. But, no species of a determinable kind could exist if that determinable kind did not exist.

Ellis distinguishes between determinate (simple) and determinable (generic) natural kinds. Determinate natural kinds are those whose members are intrinsically identical to each other. For example, the neutrons, protons, neutrinos and any other fundamental particle. However, most natural kind candidates are not as simple as the fundamental particles. Some kinds are determinable kinds, which have species; for example, the determinable kind *uranium* has its various isotopes as species (e.g. uranium-234, uranium-235, and uranium-238). Uranium atoms can be distinguished from all other kinds of atoms by the number of their protons. However, not all atoms of uranium are intrinsically identical. Two atoms of uranium can differ in their mass. Moreover, such intrinsic differences have a massive impact on how they are disposed to behave (e.g. uranium-235 is unstable and uranium-238 is more stable, both species decay in different ways). Thus, the determinable kind *uranium* has distinct species (or determinates).

According to Ellis, determinable kinds and their species share a common essence, but species have some essential properties, not common to all members of the determinable kind and it is in virtue of these properties that the species of determinable kinds are distinguished. Ellis formulates it in the following way:

(1) X is a member of the determinable kind K that has Q essentially.
(2) X is a member of a species of K that has Q essentially, but also has P essentially.

The set of properties (Q+P) constitute the real essence of the species S, whereas the property Q constitutes the essence of the members of S in K. Let K be the determinable kind *uranium*, then the atomic number 92 is the property Q that any member x of the kind K will have essentially. If x is an instance of uranium-235, then it is a species of K that has Q (the atomic number 92), essentially, but also has a different number of neutrons (235) in the nucleus to other species of the determinable kind *uranium*. Thus, the property of having 235 neutrons in its nucleus is essential to the species uranium-235. Natural kinds require a distinct ontological category, because they have essences. The essences of generic kinds existentially entail the essences of their infimic species. Chemical elements provide a paradigmatic example of determinable kinds for Ellis.

Ellis's determinable kinds can be formulated in the following way:

DET1: If K is a determinable kind, then there is some determinate kind k, which is a species of the genus K and there is one and the same set of essential properties Q, that K and k share.
DET1: If K (*uranium*) is a determinable kind, then there is some determinate kind k (uranium-235), which is a species of the genus

K and there is one and the same set of essential properties Q (atomic number 92), that K and k share.

Nevertheless, Ellis argues that some determinable kinds (e.g. plant and animal species) are not real natural kinds. They are merely clusters of similar natural kinds, which are conceptualised as being of the same kind. Ellis (2001: 170) accepts Wilkerson's (1995) view that biological species are more or less salient clusters of intrinsically similar natural kinds. Caplan (1981) is basically correct to think of species as being taxonomised in virtue of the genotypic families to which they belong. However, there is no sharp distinction between genotypic families either. Ellis (2001: 170) states:

> From the point of view of genetics, the only strict biological kinds that have more than one member are kinds consisting of genetically identical twins or clones.

Thus, *biological species* are not determinable kinds according to Ellis, because there is no set of standing intrinsic properties or structures that would distinguish the members of these kinds from all other kinds. In terms of the above formulation, there is no Q that every member of a given species has intrinsically. Biological species *qua* determinable kinds cannot be rendered determinate, rather the members of a species are themselves determinables, whose instances admit genotypic variation. Thus, for Ellis, though determinable kinds are admissible into our ontology, they should only to be given ontological significance, when the members are themselves concrete determinates.

3   Kinds of Determinable Kind

I wish to argue that there is an important distinction between different kinds of determinable kind. Some determinable natural kinds have members that are determinates (e.g. chemical elements such as *uranium* with its various isotopes as discussed in Ellis above). Other determinable kinds have members that are themselves complex determinables (e.g. *species, metals, crystals, solids, mammals, proteins, vertebrates, viruses,* and so on).

Consider the example of the chemical natural kind *metals*. There are many different members of the kind *metals*. All metals are characterized by the property of high electrical conductivity. However, the kind metals is a determinable kind, there are many determinates of the kind metal. For example, silver, gold and iron are all classified as metals. Naturally occurring iron (Fe) consists of four isotopes ^{54}Fe ^{56}Fe, ^{57}Fe and ^{58}Fe. So, the kind *iron* is itself also a determinable. Thus, the members of the determinable kind *metals* are themselves determinables.

Recall Ellis's formulation above:

> (1) Members of the determinable kind *metals* have the property of high electrical conductivity essentially.

(2) The kind *iron* (Fe) is a member of the determinable kind *metals* and it has high electrical conductivity essentially.
(3) ^{56}Fe is a determinate member of a species of the determinable kind iron Fe and it has high electrical conductivity essentially.

Thus, it would appear that determinable kinds whose members are also determinable kinds can be accommodated according to the essentialist paradigm, once the sub-determinable itself has a determinate species. Thus, Ellis can allow the following formulation of these kind of determinable kinds:

> DET2: If K is a DET2 determinable kind, then the determinate kind k is a member of the DET1 determinable sub-K, which is a sub-determinable of the determinable K and K, sub-K and k all have Q essentially.
>
> DET2: If K (*metals*) is a DET2 determinable kind, then the determinate kind k (^{56}Fe) is a member of the DET1 determinable sub-K (iron, Fe) which is a sub-determinable of the determinable K (*metals*) and K, sub-K and k all have high electrical conductivity essentially.

Nevertheless, recall that for Ellis, the direction of existential entailment must be identical to the direction of ontological dependence. However, the existence of metals does not existentially entail the existence of iron, nor indeed does it entail the existence of ^{56}Fe. Thus, the kind metal is not ontologically dependent on the kind Iron. This would seem correct since there are many different instantiations of the kind *metals* (e.g. silver and gold). Consequently, the kind *metals* is a DET2 determinable kind, which is neither ontologically dependent on its sub-determinable nor on the determinate species of this sub-determinable.

One possible response is to claim that there is some other determinable kind, under which this kind can be subsumed, but which itself satisfies either DET1 or DET2 (which, itself can be rendered determinate and which can be shown to be ontologically dependent on this determinate.) One might argue that the kind metal is ontologically dependent on there being some kind of stuff that has the property of electrical conductivity, namely there must be a natural kind (*electrical conductors*) to which all metals belong.

> (1) Members of the determinable kind *electrical conductors* have movable charged particles essentially.
> (2) The determinable kind metal is a member of the determinable kind *electrical conductor* and it has movable charged electrons essentially.
> (3) The kind Iron (Fe) is a member of the determinable kind *electrical conductor* and the determinable kind metal and it has movable charged electrons essentially.
> (4) ^{56}Fe is a member of a species of the kind *iron* and it has movable charged electrons essentially.

Importantly, in this example, the direction of ontological dependence is the same as that of existential entailment. The kind *electrical conductor* existentially entails the existence of *movable charged particles*, thus electrical conductors are ontologically dependent on moved charged particles. All metals can be subsumed under the more generic determinable *electrical conductors*. There are nevertheless, some non-metallic electrical conductors (e.g. plasmas, salt solutions). Thus, the kind *metals* is not equivalent to the kind electrical conductors. Rather, it can be subsumed by it. Thus, Ellis can argue there is a broader genus under which the kind *metals* can be subsumed, namely, the kind *electrical conductors*. It can be formulated in the following way:

> DET3: If K is a DET3 determinable kind, then the determinable K is
> *either* a sub-determinable of some DET1 determinable G, which has g as a determinate .
> *or* a DET2 determinable kind which has g as a determinate of some sub-determinate sub-G and in either case, the determinate g is an essential determinate of the determinable kind sub-K and the determinate kind k.

For example:

> DET3 If *metals* are a DET3 determinable kind, then *metals* (K) are a sub-determinable of the DET2 determinable kind *electrical conductors* (G) which has *movable charged particles* as its sub-determinate (sub-G) and *movable charged electrons* as its determinate (g) and *movable charged electrons* are also an essential determinate of the determinable kind sub-K (*iron*) and the determinate kind k (^{56}Fe).

Thus, it would appear that the essentialist could accommodate determinable kinds whose members have complex determinables. Determinable kinds must existentially entail their determinate or in the event that they do not, then there must be some higher level determinable, which subsumes them and which must itself existentially entail some determinate or subdeterminable which can itself be rendered determinate. Thus, in the above example, even though the kind *metals* does not existentially entail the kind *iron* (Fe) or any of its isotopes (e.g. ^{56}Fe). Nevertheless, there is some further determinable kind (e.g. *electrical conductors*) that can subsume it and that itself can be rendered determinate (e.g. charged electron). Thus, the existence of electrical conductors existentially entails the existence of movable charged particles and the former is ontologically dependent on the latter.

So, it would appear that the essentialist can accommodate cases of determinable kinds whose members are themselves determinables (DET3 determinable kinds). In this account, there are two commitments: the first is the necessity of existential entailment and its equivalence with ontological dependence and secondly, the hierarchical arrangement of natural kinds in subsumption relations. It is this second issue that I wish to take issue with in the next section. I will argue that simplistic nested hierarchies are too strict to accommodate determinable kinds when we consider the relationships between scientific kinds in natural science. I will suggest that crosscutting determinable kinds (DET4) require a distinct category.

4   Crosscutting Determinable Kinds

As I have argued elsewhere (2010a), crosscutting kinds present a problem for the kinds of hierarchies that we have seen above.[4] Rather crudely, the hierarchy thesis attempts to subsume lower level kinds beneath a common higher-order genus. The chief requirement for hierarchies of natural kinds is categorical distinctness. There cannot be a smooth transition from one kind to another. Moreover, if two natural kinds crosscut each other, then one must be subsumable under the other *or* there must be a common genus under which they can be subsumed.

Taking the example above, all members of the kind *iron* are members of the kind *metals* and all instances of the isotope of iron ^{56}Fe are members of the kind *iron*. Thus, according to the hierarchy thesis, ^{56}Fe can be subsumed under the kind *iron* and the kind *iron* can be subsumed under the kind *metals*. Moreover, the kind *metals* as we saw above, can be subsumed under the higher genus *electrical conductors*.

A closer examination of some examples from science reveals that this is not always the case. Consider, the biochemical kinds *proteins* and *enzymes*. Albuminoids and lipases are classified together as *proteins*. Albuminoids are water-soluble proteins. An example is serum *albumin*, which is the largest plasma protein in humans and is composed of 584 amino acids.[5] Lipases are water-soluble proteins that catalyse hydrolysis in water insoluble lipids. An example is lysosomal lipase, which is a form of lipase, which functions intracellularly in the lysosomes. Lipase is also classified as an *enzyme*. Clearly then, lipase belongs to the natural kind categories *protein* and *enzyme*. Thus, the kind categories *protein* and *enzyme* crosscut one another.

Nevertheless, proteins are not a subkind or enzymes and enzymes are not a subkind of proteins. Some enzymes are not proteins (e.g. Ribozymes are autocatalytic RNA molecules that can also be classed as *enzymes*).[6] Ribozymes

---

[4] This point has also been made by Khalidi (1998).

[5] Serum Albumin transports essential fatty acids from fat tissue to muscle tissue. It also has a role in the regulation of osmosis, and helps to transport substances (e.g. hormones) through the blood.

[6] Until the 1980s all known enzymes were thought to be proteins, until Thomas C. Cech discovered that RNA molecules could themselves catalyse chemical reactions. These catalytic RNA molecules are called Ribozymes. Until Cech's discovery, RNA (ribonucleic acid) was considered to be merely a copy of the instructions given in DNA. RNA was only a messenger that could direct protein synthesis. Cech discovered that RNA could itself fold into different shapes and in so doing

and lipases can be classified together as enzymes. Therefore, even though lipases and ribozymes can be classified together as *enzymes* and lipases and albuminoids can be classified together as *proteins*, albuminoids and ribozymes are not classified together as either *enzymes* or *proteins*, since not all enzymes are proteins and not all proteins are enzymes. This example presents a clearcut case of crosscutting natural kinds.

All the crosscutting kinds referred to are determinable kinds, whose members are themselves determinables (e.g. proteins with the members albuminoids and lipases and enzymes with the members ribozymes and lipases.) Moreover, in this example, like the example of metals above, existential entailment does not follow. Had albuminoids never existed, the category protein still would. Moreover, the existence of a protein does not entail that it is a water-soluble protein. Thus, the existence of proteins does not entail the existence of albuminoids or lipases, nor does the existence of enzymes entail the existence of ribozymes or lipases). Ontological dependence also does not follow in these cases. We appear to have a host of crosscutting determinable kinds where ontological dependence cannot be sorted out at least on the basis of those categories alone. Consider DET1 determinable kinds:

> DET1: If K is a determinable kind, then there is some determinate kind k which is a species of the genus K and there is one and the same set of essential properties Q, that K and k share.

Thus, proteins and enzymes are not DET1 determinable kinds, because the species are radically different. There is no Q that they all essentially share.

Perhaps, crosscutting kinds are best considered as DET2 determinable kinds. Recall DET2:

> DET2: If K is a DET2 determinable kind, then the determinate kind k is a member of the DET1 determinable sub-K, which is a sub-determinable of the determinable K and K, sub-K and k all have Q essentially.

Thus, according to DET2, there should be some underlying determinate kind k, which is common to both K and sub-k. For example, if enzymes are a DET2 determinable kind, then DNA, is a determinate of the sub-determinable lipases. However, there is no Q, which enzymes, lipases and DNA share. The reason why these kinds crosscut one another and cannot be rendered determinate is precisely because there is an ontological difference in their underlying structure (e.g. Proteins are composed of DNA, but only some enzymes are, some enzymes are composed of RNA (e.g. Ribozymes)). Moreover, DNA and RNA are themselves complex determinables, which are difficult to render determinate.[7] Neither DNA nor RNA can count as the underlying determinate

---

could catalyze its own biochemical reactions. This functional role was previously thought to be restricted to protein enzymes.

[7] In fact, Armstrong (1997: 66) discusses the kind DNA as precisely an example of the kind of determinable that cannot be rendered determinate. He states:

kind k, which is the essential property that all the higher-order determinable kinds share. There is no common underlying essence that all proteins and enzymes share. Crosscutting kinds are a different kind of determinable kinds that must be accommodated in our metaphysics of natural kinds.

Is there a common genus that can subsume the crosscutting categories (e.g. biomolecule)? Can crosscutting kinds be construed as DET3 kinds? A biomolecule is any organic molecule that is produced by a living organism. The kind of determinable kind that is sought is one that existentially entails some determinate. The problem with biomolecule is that it is itself a determinable whose members are complex determinables. The kind *biomolecule* includes proteins, polysaccharides, nucleic acids and metabolites. These kinds of biomolecules are all themselves determinable kinds. Equally then, these kinds are not DET3 determinable kinds, because there is no common genus that can subsume the crosscutting categories. Recall DET3:

> DET3: If K is a DET3 determinable kind, then the determinable K, is a sub-determinable of some DET1 determinable G, which has g as a determinate (or a DET2 determinable kind which has has g as a determinate of some sub-determinate sub-G) and the determinate g is an essential determinate of the determinable kind sub-K and the determinate kind k.

Thus, if the kind *proteins* is a DET3 determinable kind, then the determinable G (biomolecule) ought to subsume it and should itself be rendered determinate. The determinates of G do not share the same essential properties. Some are composed of DNA, some RNA, some monosaccharides and so on. Thus, G is itself a complex determinable. Moreover, biomolecule does not existentially entail the existence of proteins, polysaccharides, nucleic acids or metabolites. Thus, the kind *biomolecule* is not ontologically dependent on the kinds *proteins, polysaccharides, nucleic acids* or *metabolites*. The same is true if we consider *enzymes* to be a DET3 determinable kind.

Clearly then, crosscutting kinds cannot be construed as DET1, DET2 or DET3 kinds. They are a different kind of determinable kind that must be accommodated in our ontology. Crosscutting kinds might be formulated as DET4 determinable kinds in the following way:

Using once again the powerful truism that a universal must be strictly identical in each instance, it seems there is no biological structure which will serve as the universal required. The genetic structure of human DNA is perhaps the nearest thing to such a structure ... There may well be a sufficiently abstract description of that structure which is necessary and sufficient for human DNA. But, is there any reason to think that this abstract description picks out a universal?

Armstrong argues that the only reason to treat the abstract description of DNA as a universal is because it could be considered to have some irreducible causal or nomic role. However, he claims that the causal/nomic role work would be done purely by the constituent molecules that act in virtue of their determinate properties. Thus, he sees no reason to think that the kind DNA would require a distinct natural kind category.

> DET4: If K is a DET4 *crosscutting* determinable kind which over-
> laps with F, then K is neither a DET1, DET2 nor DET3 determinable
> kind and K and F overlap because there is no determinate k for K
> and there is no set of essential properties Q, which K, F and k share.

Enzymes are DET4 determinable kinds because there is no underlying determinate kind k (DNA or RNA) and there is no set of properties Q that enzymes,
proteins, ribozymes, lipases and albuminoids share. DET4 determinable kinds
crosscut precisely because there are ontological differences between their determinates, which prevent higher order subsumption. Thus, crosscutting
is ontologically significant and hierarchical subsumption in these examples
would be ontologically misleading.

Ellis would reply that DET4 determinable kinds ought to be construed as
functional kinds.[8] Functional kinds are not natural kinds, by Ellis's definitions.
Enzymes and proteins are functional kinds, for their essence (a) depends on
what the kind does, but (b) does not depend on how it does it. There is no
common causal power, and hence no common intrinsic essence involved in
the mechanism that produces functional kinds. Since, the mechanisms that
produce them have no common intrinsic essence, then they are not *sui generis*
natural kinds.

They are similar to spectral kinds as defined by Ellis (2001: 79–82). For
example, refraction and diffraction both produce spectra, and these mechanisms are natural kinds. But the mechanisms by which they produce spectra
are essentially different. Hence, *spectral producers* is a functional kind. In a
similar vein, consider the DET4 functional kind *ribozymes*, all species of ribozymes (e.g. the hairpin ribozyme and the hammerhead ribozyme) are self-
splicing RNA molecules that function as catalysts. However the mechanisms in
virtue of which they function as catalysts are essentially different. Thus, DET4
kinds such as ribozymes are functional kinds. Ellis would argue that functional
kinds, such as ribozymes and spectral producing, are not natural kinds.

However, this line of response means that most higher-level scientific
kinds will turn out not to be natural kinds at all. The hierarchical structure
provided by the essentialist will really only apply to lower level chemical and
physical kinds (e.g. elements, compounds and fundamental particles). Surely,
however, it is uncontroversial to think of kinds like *DNA* as natural kinds. We
could even provide an essentialist reading: all DNA molecules have a common
intrinsic essence, they are all double stranded macromolecules composed of
two polynucleotide chains, held together by weak thermodynamic forces. The
intrinsic causal power that all DNA molecules share is that they have the disposition to replicate. This disposition is multi-track in that DNA can encode
many different proteins, enzymes and so on.

The problem is that DNA replication can equally be construed as a functional kind. Just like spectral producing, the species of DNA that are involved

---

[8] I am grateful to Brian Ellis for suggesting this reply to me in correspondence.

in replication are intrinsically different to each other. It is differences in DNA replication and subsequent DNA folding that allows for so much biological diversity. For example, I have argued elsewhere that some non-globular proteins are intrinsically disordered in their native state. Moreover, is is this intrinsic disorder which allows them to be functionally multi-track (Tobin 2010b). Such proteins are defined solely in terms of their functional roles, because one and the same binding region can result in different biochemical kinds. If the essentialist is committed to his ontology being led by science, he must give us an account of the relationship between functional kinds and the underlying natural kinds of which they are composed.

Thus, the essentialist might respond by claiming that DET4 determinable kinds are not *sui generis* natural kinds because they do not fulfil the essentialist criteria for natural kinds; namely they have no common intrinsic essence. However, I would suggest that science informs us of a different kind of natural kind; namely DET4 determinable kinds, which ontology should reflect. Thus, either the criteria suggested by the essentialist are too stringent to accommodate some scientific kinds, or at least the essentialist should provide us with an argument for why these scientific kinds do not require an ontological account.

5   Towards An Ontology for Crosscutting Determinable Kinds

At the beginning of the paper, it was suggested that there is a consensus amongst ontologists that any ontological account of natural kinds must be informed by the putative scientific kinds found in natural science. If what I have said above is correct, then science informs us of a different kind of determinable kind; crosscutting determinable kinds (DET4), which need to be accommodated in our ontology.

The essentialist cannot accommodate DET4 determinable kinds. In his account, there are two commitments: the first is the necessity of existential entailment and its equivalence with ontological dependence and secondly, the hierarchical arrangement of natural kinds in subsumption relations. For DET4 determinable kinds, existential entailment is not equivalent to ontological dependence. DET4 determinable kinds are ontologically dependent on their determinates, but DET4 determinable kinds do not existentially entail their determinates. So, the first requirement is not satisfied for DET4 determinable kinds. Secondly, DET4 determinable kinds crosscut in ontologically significant ways and so cannot be subsumed in simple nested hierarchies. So, the second requirement is also not satisfied for DET4 determinable kinds.

In contrast, Armstrong sees no requirement for a determinable kind category. If determinable kinds supervene, then they are no ontological addition to their base. For Armstrong, the existence of the determinate entails the existence of the determinable, but the existence of the determinable does not entail the existence of the determinate. The direction of ontological dependence from the determinable to the determinate is contrary to the direction of existential entailment from the determinate to the determinable. Yet, he concludes that there are no determinable kinds.

DET4 determinable kinds present a case where the direction of ontological dependence of the determinable on the determinate is contrary to the direction of existential entailment from the determinate to the determinable. However, each determinate is insufficient to entail the determinable. (The determinates RNA or DNA are insufficient to entail the kind *enzymes*). Thus, DET4 kinds cannot be eliminated or reduced. For DET4 kinds, existential entailment is not equivalent to ontological dependence.

Furthermore, DET4 determinable kinds seem to require a distinct ontological category, because when two kinds crosscut, there is no underlying determinate kind or universal, which both determinable kinds supervene upon. The determinable kind category is required because cases of crosscutting kinds are ontologically significant. Moreover, simple nested hierarchies are prevented precisely because there are microstructural differences. Hierarchical subsumption is misleading for DET4 kinds. Our ontology must be non-hierarchical, if it is to allow for such cases of crosscutting kinds.

In conclusion, a simple distinction between determinate and determinable kinds does not capture the degree of complexity we find in scientific natural kinds. At least four types of determinable kinds must be distinguished (DET1, DET2, DET3 and DET4). Moreover, the existence of crosscutting DET4 kinds proves problematic for essentialist (Ellis) and eliminativist (Armstrong) accounts of natural kinds. Both Ellis and Armstrong agree that one should not, proceed *a priori* to posit an ontological theory and then seek to reconcile it with the scientific worldview, but that the correct ontological account of natural kinds should be scientifically informed. A closer examination of some determinables kinds (DET4 kinds) suggests that the best ontology to accommodate them is realist about natural kinds, but non-hierarchical (allowing for cases of crosscutting), one that includes natural kinds as a distinct category.[9]

[9] I am grateful to Alexander Bird, Brian Ellis and Muhammad Ali Khalidi for comments on earlier drafts of this paper. I am also grateful for helpful comments and suggestion from the audiences at 'Themes in the Metaphysics of David Armstrong' in Nottingham, February 2008 and at 'The Metaphysics of Science' conference in Nottingham in September 2009. I also wish to acknowledge the AHRC for financially supporting a period of postdoctoral research as a core researcher on the metaphysics of science project, during which this paper was written.

SCIENTIFIC KINDS WITHOUT ESSENCES

Corinne L. Bloch

1 Introduction

In the tradition of essentialism, essences are the underlying structures of reality. According to modern essentialists, they are the hidden causes of things as they appear to us; they give rise to the nature of things qua members of a kind, distinguishing the members of one kind from other kinds (see, for example, Ellis 2001).[1] According to this approach, the essence of a natural kind is the property (or subset of properties) which gives rise to the other intrinsic, non-accidental properties of its members, including their causal powers and dispositions in interactions with their environment. Some versions of essentialism contend that these essences are the most fundamental constituents which exist at a basic physical level, such as the level of atomic and subatomic particles. For example, water is essentially H_2O. Its chemical structure is causally responsible for all of the non-accidental properties and interactions of water, such as color, taste, smell, its freezing temperature under certain conditions, etc.

A central motivation for the essentialist ontology is its ability to explain the general features of reality. Brian Ellis (2001: 63) writes, 'For the questions we have to ask are: what do we have to believe in if we wish to accept the scientific world-view, and how can we best explain the presently unexplained general features of the world according to this view?' To account for the kinds that exist in the world, their hierarchical relationship, and the relations of sameness (in one or more respects) between particulars, he posits common, causally-fundamental, intrinsic properties of kinds—their metaphysical essences.

---

[1] It should be noted, though, that one can be an essentialist without subscribing to natural kinds (see Ellis 2001: 31).

I agree that we should start, as Ellis advocates, from a scientifically well-informed view of the world and work back from there. I will suggest, however, that this approach does not commit one to essentialism. I will use scientific case-studies to show that we can accept the scientific world-view and account for scientific kinds and their causally-fundamental characteristics, without the need for positing the existence of intrinsic metaphysical essences for these kinds. I will then discuss the explanatory advantages such an approach has over the view that holds that scientifically-significant kinds must have intrinsic essences.

2   A Scientifically-informed world-view

Where should we start, then? Scientists rarely talk explicitly in terms of kinds and essences. As the tradition of essentialism recognizes, however, scientists do present us with what they take to be the essential characteristics of a kind when they form scientific definitions (Ellis 2001: 35; Copi 1954; Dubs 1943).

It is important to distinguish, as Ellis does, between 'real definitions' and other types of definitions. Richard Robinson (1950: 2), for example, lists thirteen different definitions of 'definition'. In the context of theories of concepts, definitions are usually described as listing a small set of properties that are necessary and sufficient conditions for describing an instance of a concept (Laurence and Margolis 1999). However, this type of definition will not always supply us with the essential characteristics of a kind. There is often more than one set of properties that can serve as necessary and sufficient conditions for the application of a specific concept, yet we would not say that all of these sets are 'real definitions'. Unlike mere necessary and sufficient application conditions, 'real definitions' are said to convey the essences of the kind—its causally fundamental characteristics, those which make the kind what it is (Ellis 2001: 35; Copi 1954; Dubs 1943).

Unfortunately, the essentialist, whose motivation is to accommodate the scientifically-informed view of the world, can never let down his guard and be certain that such a view of the real world has indeed been reached. This is because metaphysical essences, if these exist, are often not captured by the scientists' definitions of kinds, even when those are formulated in terms of causally fundamental characteristics. The definition, unlike the metaphysical essence, is a theoretical construct, which changes as scientists learn new facts about a kind. Thus, as Ellis (2001: 35–6) and others have observed, we could find out that something that we now take to be an essential property of a kind is not really essential, and therefore should not be part of its real definition. What scientists take to be essential characteristics—along with the definitions they form to describe them—change along with their context of knowledge. If they are to find the mind-independent, eternal metaphysical essences of kinds, they can only reach such essences—and form truly 'real' definitions—at a certain end-point of scientific research. And even if scientists eventually reveal *the* underlying essence, they can never be certain that they have indeed reached that ultimate essence of things, as it is always possible that the final essence lies at a more fundamental level. Thus, while essentialism aspires to produce an ontology that conforms to the scientific world-view, we can never be sure of what is that ultimate world-view to which our ontology ought to conform.[2]

Whether or not kinds have ultimate essences, however, scientists form definitions they view as causally fundamental throughout the development of their knowledge, and not only after reaching some final level of fundamentality. These definitions map onto causal structures in the world and scientific kinds, and have a strong explanatory power. I shall present a view that accounts for the explanatory power of such definitions, which are causally-fundamental and, at the same time, contextual. Thus, the view presented here accommodates not only the scientifically-informed world-view at some final point in the development of science (or of any particular scientific enterprise), but it explains the success of the scientific practice and the explanatory powers of scientific definitions at every stage of the enquiry.

In the following section, I shall briefly outline a view arguing that the characteristics that are stated in scientific definitions throughout the progress of science do the work that essentialists ascribe to 'real definition' by virtue of the causal relations that they capture, even if they do not spell out ultimate metaphysical essences. To avoid the metaphysical commitments that come with the term 'real definitions', I will refer to these as 'scientific definitions'.[3] In Section 4, I will use two case-studies from neurobiology to provide support for the view I will put forth. I will show that scientists formulate definitions they view as expressing the most causally-fundamental characteristics of scientific kinds. I will examine the contextual nature of these definitions, and demonstrate how the suggested approach enables us to view the scientifically-informed world-view as describing real causal structures in the world at each point in the development of science, without committing to essentialism. In the final section I will discuss the various explanatory advantages the current view has, over a commitment to essentialism.[4]

---

[2]One might object that this is not the concern of the philosopher, as the philosopher need only to state that, when scientists eventually discover real essences they discover the true structure of nature. But the philosopher need not concern himself with the criteria for when such discoveries are made. However, an ontology aimed at explaining a scientifically-informed world view, must account for the explanatory successes of the kinds and relations that scientists discover throughout the progression of science, and not just at final end-points. An ontology that is focused solely on final, fundamental essences, will not account for such successes.

[3]With this term, however, I only mean to refer to definitions in terms of causally fundamental characteristics, and not other types of definitions used in science for other purposes.

[4]In order to remain ontologically neutral, I shall speak in terms of 'kinds' and not in terms of 'natural kinds'. Furthermore, I shall take, as my examples, scientific kinds that are not necessarily agreed upon by philosophers in the natural kinds tradition to be natural kinds. I will discuss the implications of my view for the ontological status of kinds in the final section.

3 Scientific definitions

3.1 Scientific definitions as contextual

Like metaphysical essences, fundamental defining characteristics describe the nature of things by pointing to real causal relations. However, they differ from metaphysical essences in two crucial respects: their contextual nature and the spectrum of causes that they capture. With respect to the contextual nature of the defining characteristics, I follow Allan Gotthelf (2012) in taking the fundamental defining characteristic of a kind to be 'that distinguishing characteristic of its units, *from among those known*, which is *known* to be responsible for (and thus explanatory of) the greatest number of other *known* distinguishing characteristics.' (emphasis added). I shall refer to this as the fundamentality condition for the defining characteristic. Similarly, Jason Rheins writes that the defining characteristic of a kind is 'the most fundamental distinguishing characteristic of a group from its nearest neighbors, not as statements of either inner, metaphysical essences nor of more-or-less freely chosen stereotypic traits.' He further writes, '[b]y 'fundamental' I intend both an epistemological and metaphysical sense. Epistemologically, I mean the criterion most capable of explaining and/or unifying the largest number of the other criteria …Metaphysically, the fundamental criterion is the one most causally influential on the most others.' Rheins (2011) argues that the defining characteristics will always depend on the current state of scientific knowledge.

Thus, while the explanatory function of the fundamental defining characteristic is derived from causal relations in reality, it is not that single, mind-independent metaphysical essence that is out there for us to grasp. As our knowledge grows and we discover new properties and new causal relations, the known characteristics that best fit the fundamentality condition may change. As we shall see in the case-studies, this brings about a change in the definition.

Scientific definitions, therefore, are contextual. The contextuality of the defining characteristics, however, does not invalidate the definition: because they explain the other distinguishing characteristics of the particulars of a kind by spelling out *real causes in the world*, within their context these characteristics are indeed fundamental, even if future scientific research would discover characteristics that are more causally-fundamental.

It is important to note that this contextuality does not imply subjectivity. My view does not imply arbitrariness, nor does it preclude the possibility of mistakes. The fundamental defining characteristic is that which is *known*—not that which is *believed*—to be responsible for the greatest number of other known distinguishing characteristics. If, for example, future scientists discover that the causal connections that were taken to exist between a defining characteristic and many other distinguishing characteristics of a kind are false, or that the defining characteristic does not apply to all the particulars belonging to the kind, then the defining characteristic is not—and has never been—valid.

3.2   Scientific definitions as capturing a broad range of causes

The second respect in which the defining characteristics differ from metaphysical essences is the range of causes that they capture. According to essentialism, the relation between the essential characteristic of a kind and its non-essential characteristics is that of basic efficient causation. The 'real definition' spells out the underlying structure that necessitates the more superficial characteristics. I will argue that the causally fundamental definitions used in science, in contrast to 'real definitions' as construed by the essentialist, express a wider range of causes that explain the other known characteristics of a kind. I suggest that the broad spectrum of causes that can be specified in a definition includes teleological as well as efficient causes.[5]

By 'teleological causation', I do not mean that future events affect past events. Here I follow the view presented by Francisco Ayala, Harry Binswanger and Larry Wright, who argued that non-conscious life-preserving processes take place because they have the end result of being life-preserving (Ayala 1970; Binswanger 1992; Wright 1976). These authors explain that the teleological nature of such processes arises from the operation of natural selection in evolution.[6] If a process proves to be beneficial and contributes to survival, it will remain in existence in future generations precisely because it enhances survival. Thus, there is no reverse causation here, in which future events influence past events.

B. F. Skinner objects that '[a] spider does not possess the elaborate behavioral repertoire with which it constructs a web because that web will enable it to capture the food it needs to survive. It possesses this behavior because similar behavior on the part of spiders in the past has enabled *them* to capture the food *they* needed to survive.' (Skinner 1953, quoted by Wright 1976: 9) Therefore, it is wrong to attribute consequence-etiology to the action of any specific agent (or function of any biological unit). When we say that a process takes place because of its end-result, however, we use general forms. When we say that a spider constructs its web because the web enables it to catch food, this description is a short-hand for saying that spiders, in general, construct webs because web-construction, in general, has beneficial results for spiders (Wright 1976: 87–9). The agents, the actions, and the consequences are being used in a general form.

---

[5]Contrast with Ellis (2002: 13), who writes, 'Aristotle's concept of final cause ... has no role in the new essentialism'. For the essentialist, the only cause specified in real definitions is the microstructural efficient one.

[6]Wright's account is not committed to natural selection. He maintains that the consequence-etiological analysis is compatible with both natural selection and intelligent design (Wright 1976: 97).

The teleological effect is a succession of efficient causes, seen from a different perspective. But the existence of an underlying mechanism for teleological processes does not diminish their reality. Rather, the mechanism is what explains how the goal is achieved. In cases involving human consciousness, this seems obvious. I am currently typing in order to prepare this paper. The physiological mechanisms involved in this action do not make my actions any less goal-directed; it merely explains how my goal is achieved. Had typing not brought about the desired result, I would not be currently engaged in this activity—the existence of the complex neuromuscular apparatus notwithstanding. The same is true for all goal-directed actions. It is a fact that had the building of webs not been an act that enhances their survival, spiders would not build them.

One might object that since I concede that there is no reverse causation, what I describe here is not teleological causation proper. If one wishes to insist that teleology only refers to 'causation from the future', it would impose no difficulty for my position. The only point crucial for present purposes is the *real causal role* that the functional characteristic (in this case—web building) has in light of natural selection. We could avoid teleological language, and instead describe the causation as a chain of past mechanical causes, but we would then have to add to the description the forces of natural selection, otherwise these past steps would lose their causal power: without natural selection, the succession of mechanical steps that won past spiders their lunch, would not cause present spiders to repeat these steps and get their own lunch. Stating the cause of the web-building by using such long descriptions instead of using teleology as shorthand would change nothing in my argument (although definitions based on such descriptions would become long and clumsy and therefore much less useful). Instead, I suggest it is much more efficient to use teleological language as shorthand but causal shorthand nonetheless.

Such teleological accounts have enormous explanatory value. The transition (which Skinner resisted) between past and future generations is what brings natural selection—without which the functional characteristic has no such trans-generational causal force—into the picture. In addition, teleological accounts enable us to differentiate between the spider that caught a fly in its web, for example, and a spider that was just lucky to have stumbled across a dead fly. While teleological explanations can be reduced to chains of mechanical causes, their explanatory value for the various components of biological systems make them appropriate features of scientific theories and, as I hope to demonstrate, of scientific definitions. Teleological accounts enable us to see that functions are essentially causal properties, where the causality is mediated by natural selection. For example, the heart is the organ that pumps blood in our blood vessels. This is a functional definition. It is also a statement of a cause in the sense I just discussed: some compartments of the heart contract, for example, *because* this action pushes the blood in our blood vessels. Functional definitions, therefore, are not inferior to definitions in terms of basic mechanical causation, nor do they serve as temporary placeholders until a mechanical cause is discovered. Definitions must describe the most causally fundamental known characteristic—but they are not limited to only one type of causation.

We have arrived here at an account of definitions quite different from 'real definitions', as viewed by the essentialist. These definitions are contextual, in that they depend on our knowledge of various properties and causal relations within the kind in question, and on our knowledge of the particulars we differentiate the kind from. In addition, they do not necessarily describe mechanical or efficient causes on the microstructural level. After reviewing two case-studies, I will argue that such scientific definitions, despite the fact that they do not contain metaphysical essences of kinds, do much of the work that is expected from 'real definitions', thus providing us with a way to account for the success of the scientific enterprise without deferring to essences.

4   Scientific definitions: Case-studies

4.1   Nerve impulse

In this section, I will use scientific case-studies to further explain and to provide support for the view I put forth above, and demonstrate how it accommodates a scientifically well-informed world view. The first case-study I will discuss is that of the nerve impulse. In his 1932 *The Mechanism of Nervous Action*, E. D. Adrian (1932: 17) defined nerve impulse as 'the change which propagates itself down the fibre and leads to the activity of other structures such as muscles or glands'. At this point, Adrian was well aware of the electrical change that belongs non-accidentally to the nerve impulse. He was also aware that the electrical change is the mechanical cause of the subsequent activity of the muscles and glands. However, he chose to exclude the electrical change from the definition. He preferred to use a definition that is mostly functional rather than confine himself to the use of mechanical causation terms. Several years later, in *The Physical Background of Perception*, Adrian (1947: 12) stated that the impulse and the electrical change are inseparable. However, he wrote (1947: 10–13):

> the wave of activity which travels down a nerve fiber, and which is always *accompanied* by an electric effect ... from *the fortunate circumstance* that the impulses are *accompanied* by small electric currents we can make records to show us what signals are passing from moment to moment along the fibres ... If the nerve-fibre can transmit some other kind of impulse without the *accompanying* electric effect our records would only tell us part of what might be

going along it. They would only tell us about *one particular type* of
nervous message.[7]

   As the text makes clear, Adrian repeatedly referred to the electrical change
as a non-essential part of the process.[8] He further made clear that, had there
been a different process that fulfils the same function, he would still classify
it as nerve impulse. What made the functional aspect an essential one for
Adrian, despite his knowledge of the electrical change? Again, according to
the view on which I am building, a definition has to pick out 'that distinguishing characteristic of [the concept's] units, *from among those known*, which is
known to be responsible for (and thus explanatory of) the greatest number of
other *known* distinguishing characteristics' (Gotthelf 2012). This indicates two
criteria for the defining characteristics: (i) they should best differentiate the instances of a concept from others, based on the current context of knowledge,
and (ii) they should give rise to the largest number of known distinguishing
characteristics of the kind. I shall now review Adrian's definition according to
these criteria, in light of the context of knowledge available to him.

(i) Differentiation and integration

In *The Mechanism of Nervous Action*, Adrian (1932: 9, 12–13) wrote, referring
to the electrical activity during nerve impulse:

> There is no doubt about the interpretation of these potential
> changes, for they are *of the same form as the much larger changes*
> which are produced when a motor nerve is stimulated electrically
> ... In any one fibre the waves are all of the same form and the mes-
> sage can only be varied by changes in the frequency and duration
> of the discharge ... *The same kind* of electrical activity is found in
> the motor nerve fibres. When a message passes down a motor fi-
> bre from the central nervous system to arouse a muscular con-
> traction we find again a rhythmic succession of potential changes
> alike in size but varying in frequency ... the [motor] movement
> developed slowly and the development goes hand in hand with
> the increase in frequency of the potential waves. Thus the effect
> produced on the motor nerve fibres by the excited nerve cell in
> the spinal cord *is like that* produced on the sensory fibre by the
> stretched muscle spindle.

   Here, Adrian placed the nerve impulse within his conceptual hierarchy,
comparing the potential changes during nerve impulse to potential changes
during muscular contraction. He stressed that both these processes belong to
the same broader class, of a type of transmission that can change in frequency

---

[7] In this and all the following neurobiological texts, emphasis is added.
[8] See also Adrian 1932: 34.

and not in size. Within this genus, Adrian aimed to differentiate the nerve impulse from the muscular contraction. Similarly, in the chapter discussing motor fibers, Adrian repeatedly compared between the electrical activity of the motor fiber and that of the nerve fiber, stressing the similarities in the mechanisms but differences in their consequences. For example, he wrote (1932: 63):

> In a muscle fibre the propagated disturbance itself seems to differ only in time relations from the impulse in a nerve fibre; it gives *the same kind of* potential change, refractory state, and so on. But it has the *further effect* of setting off the contractile mechanism of the fibre ...

Adrian did not have the scientific tools to go deeper into the mechanism of the electrical change in both cases, and so something else was needed to differentiate between them. The fundamental difference between these two processes within the context of knowledge at the time was their function, which is what Adrian included in his definition. A definition limited to terms of efficient causation would not have been able to do so, at that point in time.

(ii) Giving rise to other known characteristics

Although Adrian was aware that an electrical change was involved in the nerve impulse, he did not know the mechanism causing the electrical change, nor did he know exactly what this change contributed to the various characteristics of the process. When *The Mechanism of Nervous Action* was published in 1932, there was some evidence supporting the quite vague 'membrane hypothesis'. The theory was rejected by the time the 1947 book was published.

While not knowing much about the underlying mechanism, Adrian was aware of various characteristics of nervous transmission, many of them were explained by the characteristics in his definition. Some of these characteristics are caused, in a mechanical manner, by the self-propagating nature of the signal, mentioned in the definition. Adrian did not know, at that point, how the propagation mechanism operates. What he did know was that it operates according to the all-or-none principle, that in every region it spreads to, it has only a momentary effect followed by a refractory period, etc. This explains many of the characteristics of nerve impulse. For example, it accounts for the distance and speed in which the signal travels. Because of the refractory period, the frequency of the signal is limited. The all-or-none propagation mechanism also causes the size of the signal to be equal in all signals from the same nerve fiber. Furthermore, since the basic propagation mechanism is the same in all fibers, the signal from different areas is generally the same.

Adrian saw other characteristics of the system as caused by the function of the nerve impulse, in the teleological manner specified earlier. Since the signal from a particular fiber is the same in size, he explained that in order to lead to differentiated activity, the discrimination between different sizes of response has to be based on difference in frequency. Similarly, since the signal from different areas is basically the same, Adrian pointed out that in order to lead to differentiated activity, the discrimination between different types of sensory input has to be based on different paths. Adrian wrote that speed and distance are, of course, crucial to ensure timely responses of the animal. Last, he remarked that the cortical map is arranged in a way that enables the animal to properly respond to different sources of stimuli, with bigger cortical areas devoted to sensory regions in which the 'fine details' are important, enabling better discrimination of sensory input from such areas. Adrian's teleological language is often explicit. For example, when talking about the structure of the cortical map, he wrote (1947: 42–5):

> This difference in the scale of the map for different parts of the body is explained by the relative importance of the information coming from them …the great changes of scale in the cortical map suggest that its arrangement is *dictated by its function* and not merely because the nervous pathways preserve some kind of order on their way up to the brain.

We see that Adrian considered various structures and characteristics of the nervous system to be caused by its function. These and other characteristics known at that time were much better explained by Adrian's relatively functional definition than by specifying the electrical change, whose mechanism was still a mystery. The function of the nerve impulse and its role in the nervous system is what gives rise, in the teleological manner specified earlier, to its various attributes. Thus, Adrian's definition contained both a mechanical cause ('the change which propagates itself down the fibre') and a functional one ('leads to the activity of other structures such as muscles or glands'). Just as, for example, the signal cannot vary in size *because* of the all-or-none propagation mechanism, it varies in frequency *because* such variation contributes to its function. Adrian's definition thus has a high explanatory value of the other characteristics of nerve impulse known at the time.[9] Adding to the definition the complex issue of electrical change when so little is known about this change and its involvement in the transmission would add nothing to our grasp of the causal structure of the kind.

As science progressed, more was learned about the electrical signal, about its role in the transfer of the message in the nervous system, and about how the characteristics of the electrical change shape the various elements of the nerve response. In 1966, Bernard Katz (1966: 9) wrote: 'For well over a century it has been known that the activity of nerves and muscles is *intimately associated* with the production of electric currents. We know now that the electric signal (action potential, or spike) … *is not just a by-product, but is the essential*

---

[9] I do not mean to imply here that the various characteristics of nerve impulse that are explained by the defining characteristic can somehow be deduced from it. I will discuss this issue below.

feature of the self-propagating message, the nerve impulse'. In contemporary textbooks as well, the electrical change is identified as the nerve impulse and not as an epiphenomenon. Kandel et al. (2000: 31,150), for example, defined the signal as the action potential (AP): 'a regenerative electrical signal whose amplitude does not attenuate as it moves down the axon.' Nicholls et al. (2001: 10) stressed that the action potentials 'are *not epiphenomena*. They *are* the only signals that provide the brain with information ....'

The changing context of knowledge brought about the change in the definition. The electrical change that had been long known to be involved in transmission was now seen in all its causal glory. Today, when much more is known about the generation and propagation of APs and about how APs convey information, it has a better explanatory value. To list just a few examples, a definition in terms of AP has an explanatory value for the changes in ion concentrations in the cell, the different propagation speed in myelinated vs. non-myelinated fibers, for the effect various chemicals have on nerve cells, etc.

While the new definition is much more mechanical, the functional element has not disappeared from the definition—it is implied by the word 'signal' (other current definitions may use alternative terms, such as 'information' etc.). Even if the underlying mechanism is clear and the efficient causes illuminated, the explanatory value of teleological causes remains. Ayala (1970: 12) writes, 'A teleological explanation implies that the system under consideration is directively organized' . Indeed, the teleological cause explains many of the characteristics of the particulars subsumed under the kind in the context of the entire biological system, and thus has a broader explanatory role than efficient causes alone. Therefore, in biological systems, it is often indispensable even when all the mechanical factors contributing to the process are known.

With regards to teleological causation, two clarifications are required. First, the teleological cause (e.g., the function of the nerve impulse) does not by itself necessitate any specific feature of the system (e.g., the formation of 'biased' cortical map). This is because the same end-result might be achieved through different mechanisms under different circumstances. This multiple realizability does not diminish the causal force of the beneficial consequences. Wright addressed this issue, stressing that it is not unique to teleological causation. In the case of simple mechanical causation, a bent rail, for example, can be said to be the cause of a train accident that actually took place even though, by itself, the fact that the rail was bent did not necessitate the accident. Similarly, in teleological cases, the beneficial consequences of a process can be said to be the cause of that process, even if the specific details of the process were not necessitated by the consequences alone (Wright 1976: 35–9). That is, if a mechanism has survived through evolution *because* of the survival value of its consequences, that is sufficient grounds for attributing consequence-etiology to that mechanism. In the case-study discussed here, characteristics such as a 'biased' cortical map are caused and explained by the function of the nerve impulse, even though this function could have, in theory, been achieved in a different way, had other factors been different. Whatever other mechanisms *could have* evolved to fulfil this function, we can explain the mechanisms that *did* evolve to fulfil this function by pointing to the survival value of the consequence, in the face of natural selection.

The second point is that whether the definition specifies an efficient cause or a teleological one, the other distinguishing characteristics, as well as the causation relation between these and the defining characteristics, cannot be derived from the definition in isolation. The various characteristics of water, for example, are explained by its molecular structure, but they cannot be deduced from it without a lot of background knowledge. Similarly, knowing that the nerve-impulse is a self-propagating signal does not enable one to deduce that the signal cannot change in size—for this, one has to know more about the self-propagation mechanism. The definition only points us to some aspects of reality, but we need to have independent knowledge of that reality in order to grasp the causal relations.

4.2   The synapse

The concept of synapse was famously introduced by Sherrington in 1897: 'So far as our present knowledge goes we are led to think that the tip of a twig of the [axon's] arborescence is not continuous with but merely in contact with the substance of the dendrite or cell body on which it impinges. Such a special connection of one nerve cell with another might be called a *synapsis*.' (Sherrington 1897: 929)

Today, a synapse is often defined as 'a specialized site of functional interaction between excitable cells' (Bennett 1977). During the 20th century, as scientific knowledge accumulated, the definition of synapse went through some changes. The question, initially, was whether the properties of delay, and especially of polarity of transmission, should be part of the definition of synapse. By looking at the reasoning of scientists, we can learn something about what roles they expected definitions to have, and why they took some characteristics to be fundamental.

Angelique Arvanitaki (1942: 90) defined synapses as 'surfaces of contact (whether axosomatic, axodendritic, or axomuscular) *anatomically differentiated and functionally specialized* for the transmission of excitations in an *irreciprocal direction*'. To understand why she formulated this definition, we must look at her context of knowledge. Arvanitaki (1942: 90) differentiated synapses from what she called ephapses, which are non-specialized electrical interactions:

> We therefore propose the term 'ephapse' to designate the locus of contact or close vicinity of two active functional surfaces, whether this contact be experimental or brought about by natural means. It may be the locus of contact of two absolutely homomorphous

surfaces, for instance an axon-axon or soma-soma contact of two nerve cells. It would therefore differ from the word synapse whose meaning is narrower and designates surfaces of contact (whether axosomatic, axodendritic, or axomuscular) anatomically differentiated and functionally specialized for the transmission of the liminal excitations from one element to the following in an irreciprocal direction.

The difference between 'synapse' and 'ephapse', as seen by Arvanitaki, is not in mechanism of transmission, as she assumed both synaptic and ephaptic transmission was electrical. The difference was the differentiation and specialization that exist at the synapse, which create a unique structure that forms an attachment, or union, which do not exist in the ephapse, and the fact that synapses, unlike ephapses, are formed between heteromorphous structures, creating unidirectional action.[10]

Harry Grundfest was one of the leading researchers in the field of neurophysiology during the 20th century. He repeatedly stated that polarity of transmission is a defining characteristic of the synapse (see, for example, Kao and Grundfest 1957). His context of knowledge, however, was very different from that of Arvanitaki. The scientific community had just experienced 'the war of the soups and sparks', in which scientists debated whether synapses are chemical or electrical. In the 1950s, the war winded down, and the dominant view was that transmission at the synapse was chemical. Grundfest made sure to exclude all non-chemical interactions from the category 'synapse'. Since polarity and delay at the synapse were now associated with chemical transmission, Grundfest's view that these were fundamental defining characteristics helped him differentiate chemical synapses from electrical transmission, which he saw as 'identical with propagation of an impulse along an axon', because it was symmetrical and unpolarized.

While Grundfest viewed polarity as necessary in the definition, towards the end of the 1950s, he proposed a new, 'extended definition'—one that included electrical inexcitability of the synapse. He explained his proposal (Grundfest 1959: 150, 154):

> ...The entire group of these *distinguishing characteristics* [of synapses] appears to be *referable to a single fundamental property* of synaptic electrogenic membrane, namely that its activity is not initiated by an electrical stimulus. Thus, there arises a *profound distinction* between the conductile activity of axons or muscle fibers and the transmissional activity at synapses. The former is electrically excitable by an applied stimulus or by the internally generated local circuit of activity. The latter is electrically inexcitable and must be evoked by a specific stimulus which in

---

[10] For further details on the specialized structures of synapses, and their physiological consequences that further distinguish synapses from ephapses, see Arvanitaki (1942).

the context of synaptic structure must be a chemical excitant, or transmitter agent, released by the active presynaptic nerve fibers. The currently used definition of synapses is still essentially as it developed with Sherrington and Ramon y Cajal, a junction in contiguity between anatomically distinct cells across which activity is nevertheless transmitted, but only in one direction, from the presynaptic cell to the postsynaptic. Many other specifications are now available to *distinguish transmissional activity from conductile or ephaptic*, and these appear to *derive from the one feature*, that synaptic activity is electrically inexcitable.

...Other properties of [post-synaptic potentials] that *distinguish them from spikes* are also *referable* to this *single, fundamental difference* in their modes of excitation.

We can see that Grundfest explicitly stated that all these newly established characteristics that distinguish synapses from conductile and ephaptic transmission are derived from this single characteristic. Among these, he included the delayed and graded nature of the synaptic response (in contrast to the immediate, all-or-none nerve-trunk conduction), insensitivity of the synaptic response to field currents and to the membrane potential (in contrast to the sensitivity of nerve-trunk conduction to these factors), high sensitivity to chemicals (in contrast to the less sensitive nerve-trunk conduction), the non-propogating nature of the junctional response (in contrast to the propogating nature of nerve-trunk conduction) etc. Grundfest explained how these and other distinguishing characteristics of synaptic transmission are caused by electrical inexcitability, providing 'profound distinction' between such transmission and nerve-trunk conduction, and argued that it should therefore be included in the definition. He also argued that electrical inexcitability is more causally fundamental than polarity, which makes the new definition more causally fundamental (Grundfest 1959).

It is important to note that Grundfest thought transmission was chemical long before 1959, and he had published extensively on chemical transmission and electrical inexcitability of the synapse. However, only in 1959 did he propose to include it in the definition of synapse. Around that year, new reports showed recordings from electrotonic junctions which were polarized (Furshpan and Potter 1957; Furshpan and Potter 1959; Watanabe 1958). As a result of this new information, the old definition, which included polarization but did not specify electrical inexcitability, could no longer differentiate chemical synapses from other junctions. To maintain the differentiation, Grundfest had to revise the definition.

Gradually, in light of the growing knowledge of the functional similarities between specialized electrotonic junctions and the chemical synapses, the scientific community began to accept these junctions as synapses, and view them as belonging to the same kind, despite the differences that Grundfest emphasized.[11] John Eccles, a leading figure in the field, who in his 1961 review still proposed that the junctions between segments of septate axons should not be considered synapses because they conduct in both directions (Eccles 1961: 363; c.f. Bennett 1985), revised his definition of synapse by 1964, to 'a structure that is formed by the close apposition of neurone either with neurone or with effector cell and that is specialized for the transmission of excitation or inhibition.' Accordingly, he explicitly accepted bidirectional electrical junctions as synapses Eccles (1964: VI). The emphasis on differentiation and specialization resembles Arvanitaki's definition, with the exclusion of polarity. Commenting on her definition, Eccles (1964: VI) explained why irreciprocity should be excluded, and argued for a functional criterion:

> The further distinction was made that at the synapse transmission is irreciprocal, but this irreciprocity of transmission is now known to occur for junctional transmissions indubidably ephaptic, such as the electrical excitation of intramuscular nerve fibers by the muscle spike potential, or even between two giant axons when specially treated. On the other hand reversibility of transmission is very well developed at junctional regions that because of the criteria of *design and of function* will be classed as synapses—for example the septa of some giant axons and the electrically transmitting bridges between nerve cells or between giant axons, and with the large synapses of avian ciliary ganglia.

Michael Bennett (1977), who is one of the most significant contributors to the literature on electrical synapses, defined synapse as 'a specialized site of functional interaction between neurons'. He argued that the structure and specific mechanism are not the fundamental differentiating characteristics of the synapse, but rather its functional specializations are its defining characteristics. Bennett (1972) stressed the functional similarities among synapses and demonstrated that whether they are chemical or electrotonic—they can perform the same functions (although, admittedly, with different degrees of efficiency). Like Arvanitaki, Bennett differentiated the synapse from non-specialized forms of interneuronal communication. The issue of specialization, which highlights the teleological nature of the functional definition, is crucial, according to Bennett, for differentiating synapses from other junctions that occur between closely apposed cells without obvious specialization. Thus, Bennett (1997: 394) stated that whether some sites 'are to be considered synapses or incidental or accidental sites of interaction may become clear with greater knowledge of the developmental mechanism'. To summarize, for Bennett, the differentiating characteristic of a synapse is the specialization for a

---

[11]The changes I will now describe involve not only the formulation of a new definition, but also a change in the group of referents the concept refers to. There is much more to be said about cases of reclassification, but for my purpose here all that matters is that for a given category (which is now broader), a definition is formulized which fulfils the fundamentality condition explained above.

specific function. This specialization, in turn, is the cause of the various mechanisms that are involved in synaptic activity.

4.3   Scientific definitions—Conclusions

We can now go back to the description of definitions I gave in Section 3, and discuss in more detail the conditions for a valid scientific definition. Based on the criteria I specified, a condition for the validity of a definition is that the defining characteristic, D, is causally responsible for (most of) the other currently known differentiating characteristics of the kind.

For the condition to apply, D is not necessarily the *most* fundamental cause of these characteristics—it is possible that there is a more fundamental cause, which has not yet been discovered. Furthermore, the existence of other, currently unknown characteristics of the kind, for which D is not causally responsible, does not invalidate D as a defining characteristic under the present context of scientific knowledge. Thus, this fundamentality condition is both epistemological and metaphysical. It is epistemological in that the relevant causal relations are only those known to scientists at each point in time. It is metaphysical in that the specified causal relations must exist in the world. (Again, if a definition is based on false causal relations, then the definition is not—and never has been—valid.)

In accordance with the above condition, we saw that as context changes, scientists may replace a previously valid, defining characteristic, D, with a new defining characteristic N, for a kind K. This happens in the following cases:

1. N is discovered to be more causally fundamental than D, and so better explains the other distinguishing characteristics of K. (E.g., in the case of nerve-impulse, the change in membrane potential was discovered to be the cause of known distinguishing characteristics, such as activation of muscles. In the case of the synapse, electrical inexcitability of the synapse was taken by Grundfest to be more causally fundamental than polarity.)

2. New distinguishing characteristics for K are discovered. N is causally responsible for more characteristics in the new group of the known distinguishing characteristics than D. (E.g., in the case of nerve-impulse, the change in membrane potential was discovered to be causally responsible for the many additional distinguishing characteristics of nerve impulse. In the case of the synapse, electrical inexcitability of the synapse was taken by Grundfest to be causally responsible for more distinguishing characteristics than polarity of transmission.)

3. New knowledge leads to a change in the category to which the concept refers. In this case, a new definition is formed, that fulfils the fundamentality condition for the changed category, with its new distinguishing characteristics. (E.g., the decision to include electrical junctions under the concept of synapse led to a change in the definition of synapse. Chemical transmission, delay, etc. are no longer distinguishing characteristics for the concept of synapse. The

new, functional, definition, differentiates synapses from the new group they are differentiated from—non-specialized junctions.)

In all the above cases, the fundamentality condition no longer applies for D. Interestingly, N does not have to be a newly discovered characteristic; it can be a distinguishing characteristic that was previously known for K, but whose causal fundamentality was not known. Such was the case for the electrical change, which had been known to be a characteristic of nerve impulse, but was only included in the definition when its various causal relations to many other characteristics were elucidated (a case of type 1 above). Similarly, while Grundfest had long argued for the 'electrical inexcitability' of the synapse, he included it in the definition only when the group of the distinguishing characteristics of synapse had changed, and 'electrical inexcitablity' was taken to be responsible for that newly constructed group of characteristics (a case of type 2 above).

Instead of requiring that scientific definitions specify eternal essences, they should be viewed as cognitive tools that are meant to integrate current knowledge about scientific kinds and real causal relations in the world. Therefore, their adequacy and effectiveness should always be judged according to the knowledge of the time. Their contextual nature, along with the broad spectrum of causal connections that they convey, make definitions effective tools for integrating the constantly growing scientific knowledge. They do so by specifying those characteristics that are known to give rise to the other known characteristics of a kind, thus helping unite the particular instances under the concept and differentiate them from instances of other kinds.

The two case-studies brought here show that scientists formulate definitions to include the characteristics they take to be causally fundamental and responsible for many other distinguishing characteristics. These characteristics provide unity to the kind by (i) integrating the various distinguishing characteristics of the instances (as the common cause for many of those characteristics) and by (ii) integrating the various instances of the kind, and differentiating them from other kinds (as a common cause for the distinguishing characteristics of all the instances of the kind). The examples further show that the causality the scientists attribute to the fundamental distinguishing characteristics is not limited to mechanical causation, and can even include teleological causation.[12] Last, the above examples show that what scientists consider the most causally fundamental characteristic is contextual, depending not only on the known causal relations, but also on the ability of the characteristic to explain the other distinguishing characteristics, and thus to distinguish the kind from others at each point in time.

I further argue that these causally-fundamental, contextual definitions, do not serve as mere nominal definitions, on the path for the ultimate real defini-

---

[12] I did not attempt to exhaust the various types of causes that can be captured by the definition, which may be different in other sciences. The present point is only that the causes specified in the definition need not be restricted to mechanical causes.

tion. Rather, these scientific definitions do the work of 'real definitions', even though they do not necessarily specify metaphysical essences. First, scientific definitions specify *real causes* that are explanatory of the other differentiating characteristics of a kind; they tell us something fundamental about the nature of the kind. Second, the fundamental distinguishing characteristic provides unity to the kind in the two ways specified in the previous paragraph. Third, definitions in terms of causally fundamental characteristics emphasize that scientific kinds are not arbitrary, but justified by the fact that they share not only a variety of characteristics, but also a fundamental cause. Fourth, the contextuality of the definition does not imply subjectivity. There is a right and a wrong way to define kinds, and the right way will be useful because it best conforms to the causal relations in reality (within the context of our knowledge). Last, the defining characteristics often feature in scientific explanations about the causal structure of kinds as 'essential' and as an explanation of what the kind really is.

The case studies discussed here are both from biology. In other fields, such as physics or chemistry, the account will be much simpler, since the causal relations between the fundamental characteristic and the other distinguishing characteristics are limited to efficient causation. However, while the causal structure in such a case is much simpler, the approach to definitions would be the same. To take a hypothetical example suggested by Ellis, scientists might discover additional properties of electrons, which are not yet known, and they may judge these properties to be more causally fundamental and thus revise the definition of electron. The implication of such a case for the essentialist would be that we do not yet know what electrons really are, and our current definition is not a 'real definition' (Ellis 2001: 35–6). According to the account defended here, even if the current definition of electron does not identify ultimate metaphysical essences and will be revised later, as new facts are discovered, in the current context it is causally-fundamental and unifying in all the respects discussed above, and thus it still does the explanatory work we expect 'real definitions' to do.

In the final section, I will discuss the implications that the above description of the development of scientific definitions has for ontology, and suggest that a non-essentialist approach might have some explanatory benefits in accounting for this dynamics of science.

5   Ontological commitments and the scientifically-informed world-view

So far, I have not discussed the ontological commitments that the above view requires. I will briefly show that while the view described here does not necessarily contrast with essentialist ontology, reconciling the two comes at the cost of the explanatory benefits that motivated the essentialist to begin with. I will then outline an alternative ontological account and suggest that the alternative approach can account more fully for the scientifically-informed world-view.

5.1 Essentialism and the scientifically-informed world-view

An essentialist reader can be convinced by the above description of the roles of causally fundamental definitions in science without changing his ontological position. He could maintain that when the scientific kinds are natural kinds, they share an identical essence, which is not necessarily captured by the scientists' definitions. As science progresses, scientists come closer and closer to discovering the essences of kinds, until they are (hopefully) able to formulate 'real definitions' for these kinds. However, while the above description of contextual scientific definitions is not in itself contradictory to the doctrine of metaphysical essences, it undermines the motivation for essentialism, as essentialism does not account for the explanatory success of definitions of kinds, before reaching their ultimate essences. Similarly, an essentialist may maintain that kinds that were later reclassified (as in the case of the synapse) were not real kinds, and that only the new classification reflects natural kinds. However, this, again, does not account for the explanatory success of the 'old' kinds, and their definitions.

Furthermore, according to the essentialist, the many characteristics that synapses, for example, have in common, could be explained by their constituting a natural kind. They all share an essence, which is independent of our consciousness and responsible for their various common characteristics. If this is correct, then all synapses are identical in some respect. There is, however, no actual respect in which all synapses are identical. Their actual structure and function vary from instance to instance. Even if we narrow our survey to the sub-group of chemical synapses, the various instances would not be identical. They would have, for example, a different number of post-synaptic receptors, different sizes of synaptic cleft, etc. The metaphysical essentialist addresses this difficulty by positing that they have an identical intrinsic essence, even if the actual instantiations of synapse are not identical in any respect. Such an intrinsic essence would be displayed had the synapse been isolated from all external influences.

I did not try to provide here a metaphysical argument against the essentialist ontology—I am only concerned with its supposed explanatory advantages, since a major motivation for essentialism is the accommodation of the scientific world-view. Would the above solution assist in our understanding of this world-view? Since the synapse and its causal structure are only of interest as an interactive part of an organism, it is not clear how positing such essence would assist the neurophysiologists in understanding the causal powers of the synapse. It is further unclear what sort of causal generalizations the neurophysiologist could make based on such an intrinsic posited essence. It seems that rather than assisting in explaining the observed phenomena, it will add to the list of things to be explained.

Of course, the metaphysical essentialist might not insist that synapses have such an identical internal essence. He might reply, instead, that synapses are simply not a natural kind. But then, the metaphysical essentialist would have to concede that the causal structure that synapses share, as well as the successful inductions and generalizations scientists make about synapses, is some sort of a 'cosmic accident'. This solution, too, would not contribute to our understanding of the scientifically-informed world view.

5.2   Non-essentialist accounts and the scientifically-informed world-view

I. Scientific kinds are real, and are unified by a causally-fundamental characteristic

What alternative account would take scientific kinds as real on the one hand, without being committed to the existence of identical, intrinsic essences on the other hand? And what would it mean to say that scientific kinds are real, according to this account? The starting point is that the causal structures in the world exist independently of human knowledge or interests. Proper scientific kinds, while also dependent on human classification decisions, are nevertheless real since they reflect these independent causal structures: various instances of synapse would have many characteristics in common because they all share the same causally fundamental characteristic. That is, because all synapses are specialized sites of functional interaction between excitable cells, they have many other characteristics in common—characteristics that are explained by this specialization. Synapses would further differ from ephapses in many respects, because ephapses are not specialized for such interaction. The explanatory force of scientific definitions is grounded in the causal relations that they capture, without the need of positing the existence of underlying essences.

Importantly, my account allows one to take scientific kinds as real—within their context—not only without requiring essences, but also without requiring scientific kinds to be eternal natural kinds. Thus, even if later reclassifications bring about a change in the concept's referents, the explanatory success of the 'old' kind can still be accounted for by the causal structure that this kind reflects, and the previous context of knowledge.

The view that kinds depend both on our classification and on the causal structure of the world is similar to Richard Boyd's (see, for example, Boyd 1989; Boyd 1999). However, in the account I suggest, the particulars of the kind do not merely share imperfectly a cluster of properties. They all share the fun-

damental characteristics that give rise to these properties. Having the fundamental characteristic is critical for membership in the kind. Grundfest (1959),
for example, argued that since the junction activating the giant axon of the
squid conforms to his extended definition of synapses, it should be included
in this category. Later on, since specialization for a certain function was determined to be a fundamental characteristic for synapses, Bennett (1997) argued
that, regardless of the structural characteristics of some sites of interaction,
whether or not they should be considered as synapses depends on developmental research. We see that in contrast to Boyd's homeostatic property cluster account, in which no single property can determine kind membership, the
causally-fundamental distinguishing characteristics have a unique status for
determination of membership in the kind.[13] As explained in Section 4, this
causally-fundamental characteristic also provides the kind with unity, which
is lacking in accounts such as Boyd's.

II. A broad range of scientific kinds

The various instantiations of synapses share a causal structure that unifies
them by integrating their various distinguishing characteristics. In contrast
to the essentialist approach, however, there need not be a respect in which
the synapses are identical. This enables us to account for a broader range
of kinds than the essentialist. Instances of synapse, which share such fundamental characteristics, do not have to be *identical* in any respect to be members of a kind, not even in some intrinsic manner. Not only is one instantiation of synapse different, as a whole, from another, but the actualization of
its fundamental defining characteristic—its specialization for interaction between neurons—can be different among the various instantiations of synapses
(for example, the specialization can be instantiated in the form of different
types of post-synaptic receptors). Since many scientific kinds have causally-
fundamental, defining characteristics, but may not have an identical essence,
the current account seems to accommodate better the scientifically-informed
world-view. In addition, the broad-range of causal structures allowed under
the current view enables us to account for a larger variety of kinds than those
allowed under metaphysical essentialism.

---

[13] Two remarks should be made here. First, we must not forget that the above case-studies have
shown that the definition is not 'the final word' on classification decisions, and can change with
new knowledge. Definitions are dependent on the various distinguishing characteristics of a kind,
and change when these characteristics change. However, as long as the definition stands, it serves
as a condition for membership in the kind. Second, membership is not determined by the definition in isolation. For instance, there may be special cases in which the causally fundamental,
defining characteristic is missing, and the particular will still be classified as a member of the
kind. Such cases require a separate discussion, as they involve various considerations relating to
the broader context of the classification (see also Rheins 2011, footnote 48). For my purpose here,
all that matters is that the fundamental defining characteristic is, in the sense explained here,
'privileged', despite the additional considerations for determination of membership in a scientific
kind.

III. Kinds have hierarchical structure

So far, we have seen that the suggested approach can account for the success of scientific practices based on scientific kinds and definitions, throughout the development of science, without positing ultimate metaphysical essences. It can also account for a broader range of scientific kinds. I would now like to turn to another part of Ellis's motivation for the ontology he lays out. Ellis contends that essentialist ontology explains the hierarchical structure of kinds. Indeed, his distinction between essential, incidental and accidental characteristics is very useful for relating kinds in different levels of generalities (see Ch. 2 in Ellis 2001).

An ontology based on homeostatic property clusters, in contrast, does not seem to be able to explain the hierarchical structure of kinds. Can the causally-fundamental distinguishing characteristic help in this task? Consider, again, the case of the synapse. As we have seen, all synapses have specializations for interaction between neurons. These specializations are the cause of many characteristics which both chemical and electrical synapses have in common. Among those are the ability to synchronize activity of neurons, the ability to induce inhibitory and excitatory responses, their activity-dependent change in effectiveness, etc. (Bennett 1977, 1997). The kind 'synapse' is also divided into two sub-kinds: 'electrical synapse' and 'chemical synapse' (see Fig. 1).

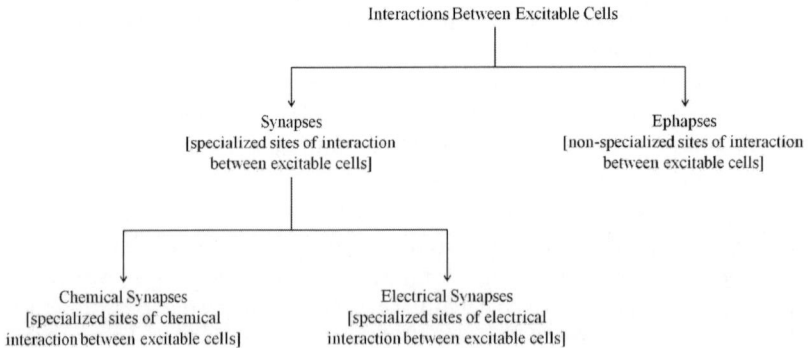

Figure 1: The hierarchical structure of kinds of interactions between excitable cells.

For the more general kind, 'synapse', the fundamental distinguishing characteristic is the specialization for interaction between neurons, and the specific mode of that interaction (chemical or electrical) is not a distinguishing characteristic. For the sub-kinds, however, the mode of interaction for which the junction is specialized becomes a fundamental differentiating characteristic, giving rise to the many, more specific characteristics that distinguish between the two sub-kinds. The fundamental distinguishing characteristic of chemical synapses—their specialization for chemical interactions—gives rise to the characteristics differentiating them from electrical synapses: they are generally unaffected by field-potential, they are slow, etc. Similarly, the characteristics differentiating electrical synapses from chemical ones are caused by their fundamental distinguishing characteristic: they are specialized for electrical interactions.

To summarize, when we go down from the more general kind 'synapse' to the sub-kinds 'electrical synapses' and 'chemical synapses', the causally-fundamental distinguishing characteristics of the more general kind (specialization for interaction between neurons) provides the broader group (the genus) within which the two sub-kinds are differentiated. Within this genus, the two sub-kinds are distinguished from each other by more specific causally-fundamental characteristics (chemical transmission vs. electrical transmission) and the various other differentiating characteristics they give rise to. Thus, we have no difficulty relating kinds in different levels of generality to each other, based on the causal structure of the world, without need for positing metaphysical essences.[14]

In this paper, I aimed to show that one need not be a metaphysical essentialist in order to acknowledge scientifically-significant kinds. I showed that scientists formulate definitions in terms of causally-fundamental, yet contextual, distinguishing characteristics. I demonstrated how taking scientific definitions as expressing explanatory characteristics that unify kinds and distinguish them from others, enables us to take seriously the scientifically informed world-view not only at the end points of scientific enterprises, but throughout the development of science. This approach accounts for scientific kinds and their causal structure, as well as for their hierarchical relationships. It provides us with a framework in which scientific advances do not pose a threat for previous knowledge, enabling us to keep our realism not only about tomorrow's scientific discoveries but about yesterday's as well. And it does all that without the need of positing metaphysical essences.[15]

---

[14]Again, this conceptual hierarchy is contextual, and may change if reclassification occurs.

[15]I thank the audience at the Metaphysics of Science conference in Melbourne in July 2009 for helpful criticism. I also thank Brian Ellis, Michela Massimi, Eva Jablonka, James Lennox, Allan Gotthelf, Patrick Mullins and an anonymous reviewer for excellent comments on previous versions of this manuscript.

REFERENCES

Adrian, E. D. 1932. *The Mechanism of Nervous Action: Electrical studies of the neurone.* Philadelphia PA: University of Pennsylvania Press.

Adrian, E. D. 1947. *The Physical Background of Perception.* Oxford: Clarendon Press.

Armstrong, D. M. 1973. Beliefs as states. In R. Toumela (Ed.), *Dispositions,* pp. 411–25. Dordrecht: Reidel.

Armstrong, D. M. 1978. *Nominalism and Realism. Universals and Scientific Realism Volume I.* Cambridge: Cambridge University Press.

Armstrong, D. M. 1983. *What is a Law of Nature?* Cambridge: Cambridge University Press.

Armstrong, D. M. 1989. *A Combinatorial Theory of Possibility.* Cambridge: Cambridge University Press. English.

Armstrong, D. M. 1997. *A World of States of Affairs.* Cambridge: Cambridge University Press.

Armstrong, D. M. 1999. Author's response. *Metascience* **8**: 85–91.

Armstrong, D. M. 2000. The causal theory of properties: Properties according to Shoemaker, Ellis and others. *Metaphysica* **1**: 5–20.

Armstrong, D. M. 2004. *Truth and Truthmakers.* Cambridge: Cambridge University Press.

Armstrong, D. M. 2005. Four disputes about properties. *Synthese* **144**: 309–20.

Armstrong, D. M., C. B. Martin, U. T. Place, and T. Crane (Ed.) 1996. *Dispositions: A debate.* London: Routledge.

Arvanitaki, A. 1942. Effects evoked in an axon by the activity of a contiguous one. *Journal of Neurophysiology* **5**: 89–108.

Ayala, F. 1970. Teleological explanations in evolutionary biology. *Philosophy of Science* **37**: 1–15.

Azzouni, J. 2004. *Deflating Existential Consequence: A case for nominalism.* Oxford: Oxford University Press.

Baez, J. 2000. What is a background-free theory? Internet publication URL = <http://math.ucr.edu/home/baez/background.html> accessed 06.06.2009.

Balaguer, M. 1998. *Platonism and Anti-Platonism in Mathematics.* New York: Oxford University Press.

Barker, S. 2009. Dispositional monism, relational constitution and quiddities. *Analysis* **69**: 242–50.

Batterman, R. 2006. Hydrodynamics versus molecular dynamics: Intertheory relations in condensed matter physics. *Philosophy of Science* **73**: 888–904.

Bennett, M. V. L. 1972. Electrical versus chemical neurotransmission. *Research Publications—Association for Research in Nervous and Mental Disease* **50**: 58–90.

Bennett, M. V. L. 1977. Electrical transmission: A functional analysis and comparison to chemical transmission. In *The Handbook of Physiology—The Nervous System I*, pp. 357–416. Washington: American Physiological Society.

Bennett, M. V. L. 1985. Nicked by occam's razor: Unitarianism in the investigation of synaptic transmission. *Biological Bulletin* **168**: 159–67.

Bennett, M. V. L. 1997. Gap junctions as electrical synapses. *Journal of Neuroctology* **26**: 349–66.

Bhaskar, R. 1975. *A Realist Theory of Science*. Leeds: Leeds Books.

Bigelow, J. 1988. *The Reality of Numbers: A physicalist's philosophy of mathematics*. Oxford: Oxford University Press.

Bigelow, J. and R. Pargetter 1988. Quantities. *Philosophical Studies* **54**: 287–304.

Bigelow, J. and R. Pargetter 1990. *Science and Necessity*. Cambridge: Cambridge University Press.

Bilson-Thompson, S. O., J. Hackett, and L. H. Kauffman 2009. Particle topology, braids and braided belts. *Journal of Mathematical Physics* **50**: 113505.

Bilson-Thompson, S. O., F. Markopoulou, and L. Smolin 2007. Quantum gravity and the standard model. *Classical and Quantum Gravity* **24**: 3975–93.

Binswanger, H. 1992. Life-based teleology and the foundations of ethics. *The Monist* **75**: 84–104.

Bird, A. 1998. Dispositions and antidotes. *Philosophical Quarterly* **48**: 227–34.

Bird, A. 2005a. Laws and essences. *Ratio* **18**: 437–61.

Bird, A. 2005b. The ultimate argument against Armstrong's contingent necessitation view of laws. *Analysis* **65**: 147–55.

Bird, A. 2005c. Unexpected a posteriori necessary laws of nature. *Australasian Journal of Philosophy* **83**: 533–48.

Bird, A. 2007a. *Nature's Metaphysics: Laws and properties*. Oxford: Oxford University Press.

Bird, A. 2007b. The regress of pure powers? *Philosophical Quarterly* **57**: 513–34.

Black, R. 2000. Against quidditism. *Australasian Journal of Philosophy* **78**: 87–104.

Blackburn, S. 1990. Filling in space. *Analysis* **50**: 62–5.

Boyd, R. 1989. What realism implies, and what it does not. *Dialectica* **43**: 5–29.

Boyd, R. 1999. Homeostasis, species, and higher taxa. In R. Wilson (Ed.), *Species: New Interdisciplinary Essays*, pp. 141–85. Cambridge MA: MIT Press.

Bradley, F. H. 1893. *Appearance and Reality*. London: Swan Sonnenschein.

Cameron, R. 2008. Truthmakers, realism, and ontology. In R. Le Poidevin (Ed.), *Being: Developments in contemporary metaphysics*, pp. 107–28. Cambridge University Press.

Caplan, A. 1981. Back to class: a note on the ontology of species. *Philosophy of Science* **48**: 130–40.

Carnap, R. 1936. Testability and meaning I. *Philosophy of Science* **3**: 491–71.

Carnap, R. 1937. Testability and meaning II. *Philosophy of Science* **4**: 1–40.

Cartwright, N. 1983. *How the Laws of Physics Lie*. Oxford: Clarendon Press.

Cartwright, N. 1989. *Nature's Capacities and Their Measurement*. Oxford: Clarendon Press.

Caves, C. M., C. A. Fuchs, and R. Schack 2007. Subjective probability and quantum certainty. *Studies in History and Philosophy of Modern Physics* **38**: 255–74.

Cheyne, C. and C. R. Pidgen 1996. Pythagorean powers or a challenge to platonism. *Australasian Journal of Philosophy* **74**: 639–45.

Chisholm, R. 1967. Identity through possible worlds. *Noûs* **1**: 1–8.

Colyvan, M. 2001. *The Indispensability of Mathematics*. Oxford: Clarendon Press.

Colyvan, M. 2007. Mathematical recreation versus mathematical knowledge. In M. Leng, A. Paseau, and M. Potter (Eds.), *Mathematical Knowledge*, pp. 109–22. Oxford: Oxford University Press.

Conger, G. 1925. The doctrine of levels. *Journal of Philosophy* **22**: 309–21.

Copi, I. M. 1954. Essence and accident. *The Journal of Philosophy* **51**: 706–19.

Davidson, D. 1980. Action, reasons, and causes. In *Essays on Actions and Events*, pp. 3–19. Oxford: Oxford University Press.

Dipert, R. 1997. The mathematial structure of the world. *Journal of Philosophy* **94**: 329–58.

Dowe, P. 2001. A counterfactual theory of prevention and 'causation' by omission. *Australasian Journal of Philosophy* **79**: 216–26.

Dresden, M. 1998. The Klopsteg Memorial Lecture: Fundamentality and numerical scales—diversity and the structure of physics. *American Journal of Physics* **66**: 468–82.

Dubs, H. 1943. Definition and its problems. *Philosophical Review* **52**: 566–77.

Dummett, M. 1991. *Frege: Philosophy of Mathematics*. London: Duckworth.

Eccles, J. C. 1961. The mechanism of synaptic transmission. *Ergebnisse der Physiologie, biologischen Chemie und experimentellen Pharmakologie* **51**: 300–430.

Eccles, J. C. 1964. *The Physiology of Synapses*. Berlin: Springer.

Ellis, B. 1968. *Basic Concepts of Measurement*. Cambridge: Cambridge University Press.

Ellis, B. 1999. The really big questions. *Metascience* **8**: 63–85.

Ellis, B. 2001. *Scientific Essentialism*. Cambridge: Cambridge University Press.

Ellis, B. 2002. *The Philosophy of Nature: A guide to the new essentialism*. Chesham: Acumen.

Ellis, B. 2005a. Physical realism. *Ratio* **18**: 371–84.

Ellis, B. 2005b. Universals, the essential problem and categorical properties. *Ratio* **18**: 462–72.

Ellis, B. 2008. Essentialism and natural kinds. In S. Psillos and M. Curd (Eds.), *The Routledge Companion to the Philosophy of Science*, pp. 139–48. Abingdon: Routledge.

Ellis, B. 2009. *The Metaphysics of Scientific Realism*. Durham: Acumen.

Ellis, B. and C. Lierse 1994. Dispositional essentialism. *Australasian Journal of Philosophy* **72**: 27–45.

Esfeld, M. and V. Lam 2008. Moderate structural realism about space-time. *Synthese* **160**: 27–46.

Field, H. 1980. *Science Without Numbers: A defence of nominalism.* Oxford: Blackwell.

Fox, J. F. 1987. Truthmakers. *Australasian Journal of Philosophy* **65**: 188–207.

Francescotti, R. 1999. How to define intrinsic properties. *Noûs* **33**: 590–609.

Franklin, J. 2009. Aristotelian realism. In A. Irvine (Ed.), *Philosophy of Mathematics*, Handbook of the Philosophy of Science (series edited by Dov Gabbay, Paul Thagard, and John Woods), pp. 1–53. North-Holland: Elsevier.

French, S. 1999. Models and mathematics in physics: The role of group theory. In J. Butterfield and C. Pagonis (Eds.), *From Physics to Philosophy*, Volume Cambridge, pp. 187–207. Cambridge University Press.

Funkhouser, E. 2006. The determinable–determinate relation. *Noûs* **40**: 548–69.

Furshpan, E. J. and D. D. Potter 1957. Mechanism of nerve-impulse transmission at a crayfish synapse. *Nature* **180**: 342–3.

Furshpan, E. J. and D. D. Potter 1959. Transmission at the giant motor synapse of the crayfish. *Journal of Physiology* **145**: 289–325.

Giaquinto, M. 2007. *Visual Thinking in Mathematics.* Oxford: Clarendon Press.

Gnassounou, B. and M. Kistler 2007. Introduction. In M. Kistler and B. Gnassounou (Eds.), *Dispositions and Causal Dispositions*, pp. 1–40. Aldershot: Ashgate.

Goodman, N. 1955. *Fact, Fiction, and Forecast.* Cambridge MA: Harvard University Press.

Gotthelf, A. 2012. Ayn Rand on concepts: Rethinking abstraction and essence. In A. Gotthelf and J. G. Lennox (Eds.), *Concepts and Their Role in Knowledge: Reflections on Objectivist Epistemology*. Pittsburgh: University of Pittsburgh Press.

Gribbin, J. 1998. *The Search for Superstrings, Symmetry, and the Theory of Everything.* Boston MA: Little, Brown and Co.

Grundfest, H. 1959. Synaptic and ephaptic transmission. In J. Field (Ed.), *Handbook of Physiology*, Volume 1, pp. 147–97. American Physiological Society.

Handfield, T. 2008. Unfinkable dispositions. *Synthese* **160**: 297–308.

Harré, R. 1970. *The Principles of Scientific Thinking.* London: Macmillan.

Harré, R. 2001. Active powers and powerful actors. *Philosophy* **48**: 91–109.

Harré, R. and E. H. Madden 1975. *Causal Powers: A theory of natural necessity.* Oxford: Blackwell.

Hartle, J. B. 2003. *Gravity: An introduction to Einstein's General Relativity.* San Francisco CA: Addison Wesley.

Hawthorne, J. 1996. Mathematical instrumentalism meets the conjunction objection. *Journal of Philosophical Logic* **25**: 363–97.

Heck, R. 2000. Syntactic reductionism. *Philosophia Mathematica* **8**: 124–49.

Heil, J. 2003. *From an Ontological Point of View.* Oxford: Oxford University Press.

Heil, J. 2005. Kinds and essences. *Ratio* **18**: 405–19.

Heil, J. 2007. Reply to Sharon Ford. In G. Romano (Ed.), *Symposium on From an Ontological Point of View by John Heil.* <http://www.swif.uniba.it/lei/mind/swifpmr.htm>: *SWIF Philosophy of Mind Review* vol. 6.

Hitchcock, C. 2007. What's wrong with neuron diagrams? In M. O. J. K. Campbell and H. Silverstein (Eds.), *Causation and Explanation*, Volume 4 of *Topics in Contemporary Philosophy*, pp. 69–92. Cambridge, MA: MIT Press.

Holmes, M. 1995. *Introduction to Perturbation Methods.* New York: Springer.

Horgan, T. and M. Potrč 2008. *Austere Realism: Contextual semantics meets minimal ontology.* Cambridge MA: MIT Press.

Hume, D. 1740. Abstract of a treatise of human nature. In P. Millican (Ed.), *An Enquiry Concerning Human Understanding* (2007 ed.)., pp. 133–45. Oxford: Oxford University Press.

Joachim, H. 1906. *The Nature of Truth: An Essay.* Oxford: Clarendon Press.

Johnson, W. E. 1964. *Logic*, Volume 1. New York: Dover Publications.

Johnston, M. 1992. How to speak of the colors. *Philosophical Studies* **68**: 221–63.

Kandel, E. R., J. H. Schwartz, and T. M. Jessell 2000. *Principles of Neural Science.* New York NY: McGraw–Hill Health Professions Division.

Kao, C. Y. and H. Grundfest 1957. Postsynaptic electrogenesis in septate giant axons. I. Earthworm median giant axon. *Journal of Neurophysiology* **20**: 553–73.

Katz, B. 1966. *Nerve, Muscle and Synapse.* New York NY: McGraw-Hill.

Ketland, J. 2011. Nominalistic adequacy. *Proceedings of the Aristotelian Society* (paper delivered on 10 January 2011).

Khalidi, M. 1998. Natural kinds and crosscutting categories. *Journal of Philosophy* **95**: 33–50.

Kim, J. 1998. Events as property exemplifications. In C. Macdonald and S. Laurence (Eds.), *Contemporary Readings in the Foundations of Metaphysics*, pp. 310–26. Oxford: Blackwell.

Kistler, M. 2007. The causal efficacy of macroscopic powerful properties. In M. Kistler and B. Gnassounou (Eds.), *Dispositions and Causal Dispositions*, pp. 103–32. Ashgate.

Kribs, D. W. and F. Markopoulou 2005. Geometry from quantum particles. <http://arxiv.org/pdf/gr-qc/0510052v1> (retrieved 11 Oct 2005).

Ladyman, J. and D. Ross 2007. *Every Thing Must Go: Metaphysics naturalized*. Oxford: Oxford University Press.

Langton, R. and D. Lewis 1998. Defining 'intrinsic'. *Philosophy and Phenomenological Research* **58**: 333–45.

Laurence, S. and E. Margolis 1999. Concepts and cognitive science. In E. Margolis and S. Laurence (Eds.), *Concepts: Core Readings*, pp. 3–81. Cambridge MA: MIT Press.

Leng, M. 2002. What's wrong with indispensability? *Synthese* **131**: 395–417.

Leng, M. 2005a. Mathematical explanation. In C. Cellucci and D. Gillies (Eds.), *Mathematical Reasoning and Heuristics*, pp. 167–89. London: King's College Publications.

Leng, M. 2005b. Platonism and anti-platonism: Why worry? *International Studies in the Philosophy of Science* **19**: 65–84.

Lewis, D. K. 1973a. Causation. *Journal of Philosophy* **70**: 556–67.

Lewis, D. K. 1973b. *Counterfactuals*. Oxford: Blackwell.

Lewis, D. K. 1983. Extrinsic properties. *Philosophical Studies* **44**: 197–200.

Lewis, D. K. 1994. Humean supervenience debugged. *Mind* **103**: 473–90.

Lewis, D. K. 2001. Truthmaking and difference-making. *Noûs* **35**: 602–15.

Lewis, D. K. 2009. Ramseyan humility. In D. Braddon-Mitchell and R. Nola (Eds.), *Conceptual Analysis and Philosophical Naturalism*. Cambridge MA: MIT Press.

Locke, J. 1690. *An Essay Concerning Human Understanding* (1964, A. D. Woozley ed.). London: Fontana.

Lombard, L. 1986. *Events: A metaphysical study*. London: Routledge and Kegan Paul.

Lowe, E. J. 2006. *The Four-Category Ontology: A metaphysical foundation for natural science*. Oxford: Oxford University Press.

Lyon, A. forthcoming. Mathematical explanations of empirical facts, and mathematical realism. *Australasian Journal of Philosophy*.

Madden, E. H. 1972. Discussion: R. Harré's *The Principles of Scientific Thinking*. *Southern Journal of Philosophy* **10**: 23–32.

Martin, C. B. 1993. Power for realists. In J. Bacon, K. Campbell, and L. Reinhardt (Eds.), *Ontology, Causality and Mind*, pp. 175–86. Cambridge: Cambridge University Press.

Martin, C. B. 1997. On the need for properties: The road to pythagoreanism and back. *Synthese* **112**: 193–231.

Martin, C. B. 2008. *The Mind in Nature*. Oxford: Oxford University Press.

Martin, C. B. and K. Pfeifer 1986. Intentionality and the non-psychological. *Philosophy and Phenomenological Research* **46**: 531–54.

McKitrick, J. 2003a. The bare metaphysical possibility of bare dispositions. *Philosophy and Phenomenological Research* **66**: 349–69.

McKitrick, J. 2003b. A case for extrinsic dispositions. *Australasian Journal of Philosophy* **81**: 155–74.

Mellor, D. H. 2000. The semantics and ontology of dispositions. *Mind* **109**: 757–780.

Mill, J. S. 1843. *A System of Logic, Ratiocinative and Inductive*. London: Parker.

Molnar, G. 1999. Are dispositions reducible? *Philosophical Quarterly* **49**: 1–16.

Molnar, G. 2003. *Powers: A study in metaphysics*. Oxford: Oxford University Press.

Moore, G. E. 1919. External and internal relations. *Proceedings of the Aristotelian Society* **20**: 40–62.

Mumford, S. 1998. *Dispositions*. Oxford: Oxford University Press.

Mumford, S. 1999. Intentionality and the physical: A new theory of disposition ascription. *Philosophical Quarterly* **49**: 215–25.

Mumford, S. 2004. *Laws in Nature*. London: Routledge.

Mumford, S. 2006. The ungrounded argument. *Synthese* **149**: 471–89.

Mumford, S. 2007. *David Armstrong*. Chesham: Acumen.

Mumford, S. 2008. Powers, dispositions, properties: or a causal realist manifesto. In R. Groff (Ed.), *Revitalizing Causality: Realism about causality in philosophy and social science*, pp. 139–51. London: Routledge.

Mumford, S. 2009. Passing powers around. *The Monist* **92**: 94–111.

Mumford, S. and R. Anjum 2009. Double prevention and powers. *Journal of Critical Realism* **8**: 277–93.

Mumford, S. and R. L. Anjum 2010. A powerful theory of causation. In A. Marmadoro (Ed.), *The Metaphysics of Powers: Their grounding and their manifestations*, pp. 143–59. London: Routledge.

Mumford, S. and R. L. Anjum 2011. Dispositional modality. In G. F. Gethmann (Ed.), *Lebenswelt und Wissenschaft, Deutsches Jarhbuch Philosophie 2*. Hamburg: Meiner.

Newstead, A. 2001. Aristotle and modern mathematical theories of the continuum. In D. Sfendoni-Mentzou, J. Hattiangadi, and D. Johnson (Eds.), *Aristotle and Contemporary Science II*, pp. 113–29. Frankfurt: Peter Lang.

Newstead, A. and J. Franklin 2008. On the reality of the continuum. *Philosophy* **83**: 117–27.

Newstead, A. and J. Franklin 2010. The epistemology of geometry I: The problem of exactness. In W. Christensen, E. Schier, and J. Sutton (Eds.), *Proceedings of the 9th Conference of the Australasian Society for Cognitive Science*, Sydney, pp. 254–60. Macquarie Centre for Cognitive Science.

Nicholls, J. G., R. A. Martin, B. G. Wallace, and P. A. Fuchs 2001. *From Neuron to Brain*. Sunderland MA: Sinauer Associates.

Oddie, G. 1982. Armstrong on the eleatic principle and abstract entities. *Philosophical Studies* **41**: 285–95.

Oppenheim, P. and H. Putnam 1958. Unity of science as a working hypothesis. In H. Feigl, M. Scriven, and G. Maxwell (Eds.), *Concepts, Theories, and the Mind-Body Problem*, Volume 2 of *Minnesota Studies in the Philosophy of*

Science, pp. 3–36. Minneapolis MN: University of Minnesota Press.

Parsons, C. 1986. Quine on the philosophy of mathematics. In L. Hahn and P. Schilpp (Eds.), *The Philosophy of W. V. Quine*, pp. 396–403. La Salle IL: Open Court.

Pettigrew, R. 2009. Aristotle on the subject matter of geometry. *Phronesis* **54**: 239–60.

Pincock, C. 2007. A role for mathematics in the physical sciences. *Noûs* **41**: 253–75.

Place, U. T. 1956. Is consciousness a brain process? *British Journal of Psychology* **47**: 44–50.

Place, U. T. 1996. Intentionality as the mark of the dispositional. *Dialectica* **50**: 91–120.

Place, U. T. 1999. Intentionality and the physical: A reply to Mumford. *Philosophical Quarterly* **49**: 225–31.

Popper, K. 1957. The propensity interpretation of the calculus of probability, and the quantum theory. In S. Körner (Ed.), *Observation and Interpretation*, pp. 65–70. London: Butterworth.

Prior, E. W. 1985. *Dispositions*. Aberdeen: Aberdeen University Press.

Prior, E. W., R. Pargetter, and F. Jackson 1982. Three theses about dispositions. *American Philosophical Quarterly* **19**: 251–57.

Psillos, S. 1999. *Scientific Realism: How science tracks truth*. London: Routledge.

Psillos, S. 2006a. The structure, the whole structure and nothing but the structure? *Philosophy of Science* **73**: 560—70.

Psillos, S. 2006b. What do powers do when they are not manifested? *Philosophy and Phenomenological Research* **72**: 137–55.

Psillos, S. 2011. Living with the abstract: Realism and models. *Synthese* **180**: 3–17.

Putnam, H. 1971. *Philosophy of Logic*. New York NY: Harper and Row.

Quine, W. V. 1960. *Word and Object*. Cambridge MA: MIT Press.

Quine, W. V. 1966. Necessary truth. In *The Ways of Paradox and Other Essays*, pp. 68–76. Cambridge MA: Harvard University Press.

Quine, W. V. 1971. *The Roots of Reference*. La Salle IL: Open Court.

Quine, W. V. 1980. *From a Logical Point of View* (Second Revised Edit. ed.). Cambridge MA: Harvard University Press.

Redhead, M. 1975. Symmetry in intertheory relations. *Synthese* **32**: 77–112.

Redhead, M. 1982. Quantum field theory for philosophers. *PSA 1982: Proceedings of the Biennial Meeting of the Philosophy of Science Association*: 57–99.

Resnik, M. 1985. How nominalist is Hartry Field's nominalism? *Philosophical Studies* **47**: 163–81.

Rheins, J. G. 2011. Similarity and species concepts. In M. O. J. Campbell and M. Slater (Eds.), *Carving Nature at its Joints: Topics in Contemporary Philosophy*, Volume 8. Cambridge MA: MIT Press.

Robinson, R. 1950. *Definition*. Oxford: Clarendon Press.

Rosen, G. 2001. Nominalism, naturalism, epistemic relativism. *Philosophical Perspectives* **15**: 69–91.

Rosen, G. and C. Dorr 2002. Composition as a fiction. In R. M. Gale (Ed.), *Blackwell Guide to Metaphysics*. Oxford: Blackwell.

Rovelli, C. 1997. Halfway through the woods: Contemporary research on space and time. In J. Earman and J. Norton (Eds.), *The Cosmos of Science: Essays of Exploration*, pp. 180–223. Pittsburgh PA: University of Pittsburgh Press.

Rueger, A. 2006. Functional reduction and emergence in the physical sciences. *Synthese* **151**: 335–46.

Rueger, A. and P. McGivern 2010. Hierarchies and levels of reality. *Synthese* **176**: 379–97.

Russell, B. 1910. *Philosophical Essays*. London: Longmans, Green and Co.

Russell, B. 1948. *Human Knowledge: Its scope and limits*. London: George Allen and Unwin.

Ryle, G. 1949. *The Concept of Mind*. London: Hutchinson.

Saatsi, J. 2007. Living in harmony: Nominalism and the explanationist argument for realism. *International Studies in the Philosophy of Science* **21**: 19–33.

Schaffer, J. 2000. Trumping preemption. *The Journal of Philosophy* **97**: 165–181.

Schaffer, J. 2003a. Is there a fundamental level? *Noûs* **37**: 498–517.

Schaffer, J. 2003b. The problem of free mass: Must properties cluster? *Philosophy and Phenomenological Research* **66**: 125–38.

Schaffer, J. 2007. Monism. In E. N. Zalta (Ed.), *The Stanford Encyclopedia of Philosophy (Fall 2008 edition)* (URL = <http://plato.stanford.edu/archives/fall2008/entries/monism/> ed.).

Schaffer, J. 2008. Truthmaker commitments. *Philosophical Studies* **141**: 7–19.

Schaffer, J. 2009. On what grounds what. In D. Chalmers, D. Manley, and R. Wasserman (Eds.), *Metametaphysics*, pp. 357–83. Oxford: Oxford University Press.

Schaffer, J. 2010a. The least discerning and most promiscuous truthmaker. *Philosophical Quarterly* **60**: 307–24.

Schaffer, J. 2010b. Monism: The priority of the whole. *Philosophical Review* **119**: 31–76.

Shapiro, S. 2000. *Thinking about Mathematics*. Oxford: Oxford University Press.

Sherrington, C. S. 1897. In M. Foster (Ed.), *A Textbook of Physiology, Part Three: The Central Nervous System* (7 ed.). London: MacMillan.

Shoemaker, S. 1980. Causality and properties. In P. van Inwagen (Ed.), *Time and Cause*, pp. 109–35. Dordrecht: Reidel.

Skinner, B. F. 1953. *Science and human behavior*. New York NY: Macmillan.

Smart, J. J. C. 1963. *Philosophy and Scientific Realism*. London: Routledge and Kegan Paul.

Smolin, L. 1997. The future of spin networks. <http://arxiv.org/pdf/gr-qc/9702030> (retrieved 17 Feb 1997).

Smolin, L. 2000. *Three Roads to Quantum Gravity*. London: Phoenix.

Smolin, L. 2006. *The Trouble with Physics: The rise of string theory, the fall of a science and what comes next*. Harmondsworth: Penguin.

Sober, E. 1993. Mathematics and indispensability. *Philosophical Review* **102**: 35–57.

Sperling, G. 1960. The information available in brief visual representations. *Psychological Monographs: General and applied* **74**: 1–29.

Swinburne, R. 1980. Comments and replies: The Shoemaker–Swinburne session. In L. J. Cohen and M. Hesse (Eds.), *Applications of Inductive Logic*, pp. 321–32. Oxford: Clarendon Press.

Teller, P. 1982. Comments on the papers of Cushing and Redhead: "Models, high-energy theoretical physics and realism" and "Quantum field theory for philosophers". *PSA 1982: Proceedings of the Biennial Meeting of the Philosophy of Science Association*: 100–111.

Teller, P. 1995. *An Interpretative Introduction to Quantum Field Theory*. Princeton NJ: Princeton University Press.

Tobin, E. 2010a. Crosscutting natural kinds and the hierarchy thesis. In H. Beebee and N. Leary (Eds.), *The Semantics and Metaphysics of Natural Kinds*, pp. 179–91. Abingdon: Routledge.

Tobin, E. 2010b. Microstructuralism and macromolecules: the case of moonlighting proteins. *Foundations of Chemistry* **12**: 41–54.

van Fraassen, B. 1980. *The Scientific Image*. Oxford: Clarendon Press.

van Fraassen, B. 2006. Representation: The problem for structuralism. *Philosophy of Science* **73**: 536–47.

Watanabe, A. 1958. The interaction of electrical activity among neurons of lobster cardiac ganglion. *Japanese Journal of Physiology* **8**: 305–18.

Weinberg, S. 1993. *Dreams of a Final Theory: The Search for the Fundamental Laws of Nature*. London: Vintage.

Wetzel, L. 2009. *Types and Tokens: On abstract objects*. Cambridge MA: The MIT Press.

Wilkerson, T. 1995. *Natural Kinds*. Avebury: Ashgate.

Williams, N. 2005. Static and dynamic dispositions. *Synthese* **146**: 303–24.

Williams, N. 2009. The ungrounded argument is unfounded: a response to Mumford. *Synthese* **170**: 7–19.

Wolfram, S. 2002. *A New Kind of Science*. Champaign IL: Wolfram Media.

Wright, L. 1976. *Teleological Explanations: An etiological analysis of goals and functions*. Berkeley CA: University of California Press.

Zeilberger, D. 2004. "Real" analysis is a degenerate case of discrete analysis. In B. Aulbach, S. Elaydi, and G. Ladas (Eds.), *New progress in difference equations*, Volume 7 of *Proceedings of the International Conference on Differential Equations and Applications*. London: Taylor and Francis.

INDEX